ARTE
THE CRUCIBLE

RISK AND REWARD

GRAY SUTTON

OLD SHRIMP ROAD
— PRESS —

Key West, Florida

Book Cover by Old Shrimp Road Press
(AI-assisted using Microsoft Designer)

ISBN 978-1-971363-05-9 (Paperback)
ISBN 978-1-971363-04-2 (eBook)

1st edition 2026

Artemis II Mission Patch
(Photo courtesy of NASA)

Artemis II Mission Timeline At a Glance

Core Stage Cutoff	~8 Minutes	RS-25 engines shut down; upper stage ignites
First Apogee Burn	~50 Minutes	Upper stage raises orbit to ~1,200 nautical miles
Second Apogee Burn	~2 Hours	23.5-hour elliptical orbit reaching ~38,000 miles
Proximity Ops Start	3 Hours 24 Min	Orion separates; Glover flies around upper stage
Proximity Ops End	~5 Hours 30 Min	Manual flying concludes; autopilot resumes
Perigee Raise Burn	~16 Hours	Orbit circularized for Moon burn
Translunar Injection	~28 Hours	18-minute burn sends Orion toward the Moon
Lunar Closest Approach	~4 Days	Orion passes ~7,400 km above lunar far side
Earth Return Prep	~8 Days	Crew prepares for return journey to Earth
Splashdown	~10 Days	Capsule reenters, parachutes into Pacific Ocean.

Artemis II Mission Timeline
(Infographic by MS Designer)

Following the Mission

Official Sources

- **NASA Artemis website:**
  nasa.gov/artemis
  Mission updates, livestreams,
  and press releases.

- **NASA TV:**
  Cable, satellite,
  streaming apps
  & nasa.gov/nasatv

- **@NASA &**
  @NASAArtemis
  on Social Media
  Real-time updates and commentary

- **Artemis Blog:**
  blogs.nasa.gov/artemis
  Detailed mission logs

What to Expect

- **Live coverage**
  starts T-minus
  4 hours

- **Expert guests**
  and mission explainers

- **Internal Orion**
  camera views
  (when possible)

- **Behind-the-scenes**
  & crew updates
  on social media

- **Tracking Footage**
  from ground stations
  & aircraft

- **Post-launch commentary**
  through orbit insertion

If You're Watching in Person

- **Kennedy Space Center**
  Visitor Complex
  Closest public access
  (tickets required)

- **Space Coast Beaches**
  10-20 miles away
  (Cocoa Beach, Titusville)

- **Sound delay:**
  ~50 sec between
  launch & sound

- **Bring sun protection,**
  water & patience

- **Arrive early**—roads
  & parking fill up fast

"Following the Mission" Infographic
(Infographic by MS Designer)

For Lydo, who dreams of the stars.
—Papa

Contents

Foreword

The beach felt empty that dark November night in 2022. A handful of us had gathered on the sand near Cape Canaveral, perhaps an hour before the scheduled launch of Artemis I, waiting in darkness under a moonless sky. The absence of the Moon felt oddly appropriate—we were here to watch humanity's first step back toward that distant world after half a century away.

Satellite Beach, November 2022.
(Photo: Sutton)

I've visited the Cape many times over the years. I've watched shuttles lift off, seen dozens of rockets climb into Florida's brilliant blue afternoons, felt that peculiar combination of pride and wonder that spaceflight inspires. But this was different. This wasn't a satellite deployment or a resupply run to the International Space Station. This was the Space Launch System—the most powerful rocket America had built since Saturn V—carrying an uncrewed Orion capsule on a test flight that would determine whether humans could return to deep space.

The launch window opened and closed. Technical holds. Weather concerns. The usual complications that attend ambitious undertakings. Around me, people murmured and checked their phones and speculated about whether we'd see anything tonight. The delay stretched. The beach felt desolate despite the scattered crowd. I lay back on the sand and, exhausted from the day's travel and the late hour, fell asleep.

I woke to the ground rumbling.

Night had become day. Seven miles away, eight point eight million pounds of thrust was overcoming gravity and inertia, accelerating a tower of metal and propellant and ambition skyward. The sound arrived as a physical thing—not just noise but pressure, a wave that moved through the sand and into my chest. People were cheering. Some were crying. I felt, in that moment, like a child—overcome with an emotion I struggle even now to articulate fully. It was joy, certainly. Pride. But also something deeper: a recognition that humans still attempt the difficult, still venture toward the unknown, still build machines capable of carrying us beyond the safety of our home world.

That moment—lying on a dark beach, feeling the earth shake, watching fire climb into the night—is part of why this book exists.

The technical story of Artemis II matters enormously. The engineering decisions, the calculated risks, the systems that must work flawlessly or the mission fails—these deserve rigorous examination. You'll find that examination here: the heat shield crisis that forced NASA to choose between delay and an untested reentry profile, the proximity operations demonstration that validates manual flight for future Gateway docking, the life-support systems that must sustain four astronauts for ten days in deep space. This book takes those technical realities seriously because the people building and flying these systems do.

But the human story matters just as much. Space exploration exists at the intersection of technical capability and human aspiration. We build rockets because we can, but we point them toward the stars because we must. The urge to explore, to understand, to venture beyond familiar horizons—this is as fundamental to our species as language or tool-making. When we abandon exploration, something essential atrophies.

The space program validates both dimensions. Technically, it drives innovation that cascades into medicine, communications, materials science, and countless other fields. The integrated circuits that power your phone, the water filtration systems in developing nations, the weather satellites that predict hurricanes—these are dividends from investments in space technology. Artemis will generate similar returns: advances in life support, autonomous operations, radiation shielding, and in-situ resource utilization that have applications far beyond the Moon.

But the human dimension may matter more. When Christina Koch monitors life support aboard Orion, she demonstrates that technical expertise has no gender barriers. When Victor Glover pilots the spacecraft manually during proximity operations, he shows young people of color that this frontier belongs to them too. When Jeremy Hansen represents Canada's partnership in the mission, he affirms that exploration transcends national boundaries. And when all four crew members venture farther from Earth than any humans in more than fifty years, they prove that ambitious goals remain achievable despite polarization, fiscal constraints, and short attention spans.

The ground rumbled that November night because thousands of engineers, technicians, and mission planners committed years of their lives to making it happen. The light that turned night into day represented not just chemical energy but human determination. Whether Artemis leads to sustained lunar exploration or becomes another brief symbolic gesture depends on choices we make collectively in the years ahead—choices about priorities, resources, and what kind of future we believe is worth building toward.

This book examines what comes next. It explains the mission that determines whether the fire I watched climb into that Florida night was the beginning of something enduring or simply a spectacular but temporary flare. I hope it helps you understand what's at stake when Wiseman, Glover, Koch, and Hansen strap into Orion and commit to the crucible that validates whether we're truly returning to deep space—or just visiting one more time.

—Gray Sutton
Melbourne, Florida

January 2026

Space Launch System (SLS) rocket and Orion spacecraft at Launch Pad 39B - March 18, 2022.
(Photo courtesy of NASA/Kim Shiflett)

CHAPTER ONE

Artemis

More than half a century has passed since humans last left low Earth orbit. In December 1972, the Apollo 17 crew departed the Moon's surface and humanity's relationship with deep space entered a long pause. The International Space Station circles Earth at an altitude where crews can be resupplied within days and evacuated within hours. Satellites and telescopes operate in the relative safety of Earth's orbit. But beyond that protective zone—past the Van Allen radiation belts, into the realm where Earth appears as a distant marble—no human has ventured since the last Apollo astronauts came home.

Artemis is NASA's program to end that absence. Named after the twin sister of Apollo in Greek mythology, it aims to return humans to the Moon and establish a permanent presence there. Unlike Apollo, which planted flags and collected rocks during brief visits, Artemis envisions astronauts living and working on the lunar surface for weeks at a time, building infrastructure, conducting science, and learning to extract resources from the Moon itself. The program includes a new heavy-lift rocket, a deep-space crew capsule, a small space station in lunar orbit, and commercial landers that will ferry astronauts to the surface. It's not a single mission but a campaign of flights building toward sustained human presence.

The stakes are extraordinary. Each Artemis mission costs approximately $4.1 billion—a figure that invites scrutiny and demands results. Artemis II, scheduled for 2026, must validate that the Space Launch System and Orion spacecraft can safely carry crews to lunar distance and back. If systems fail or perform poorly, the political consequences could cascade into program cancellation, ceding deep-space leadership to nations like China at precisely the moment when lunar competition intensifies. Success is not guaranteed, but the attempt itself signals that the United States and its partners believe exploration beyond low Earth orbit remains essential to national and scientific priorities.

Why does this matter now? Three forces are converging: First, technology has matured to the point where sustainable lunar operations are achievable—advanced life support, in-situ resource utilization, and commercial partnerships can make long-duration missions economically viable in ways they weren't during Apollo. Second, international competition is intensifying. China has announced plans to land taikonauts near the lunar south pole by 2030, and nations around the world are developing capabilities that could reshape who controls access to lunar resources. Third, the Moon serves as an essential stepping stone to Mars. Every system that must function reliably for a multi-year Mars mission—radiation protection, closed-loop life support, autonomous operations—can be tested and refined on the Moon, where help from Earth is only three days away instead of eight months.

The Artemis program matters because it's a test of whether the United States and its partners can still execute ambitious, complex engineering projects in an era of short attention spans and fiscal scrutiny. It matters because the decisions made now about how to share the Moon—who can extract water, how traffic will be managed, what heritage sites will be protected—will establish precedents for all future exploration. And it matters because representation matters: the diverse crews flying these missions signal that space is no longer the exclusive domain of a narrow demographic.

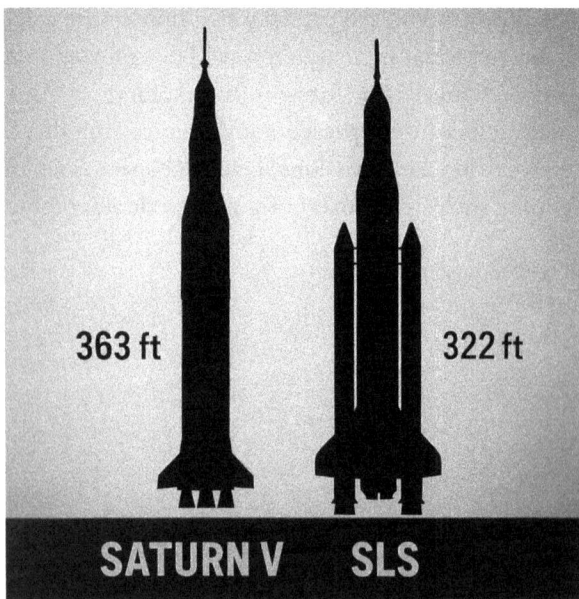

363 ft 322 ft

SATURN V SLS

Apollo vs. Artemis Comparison Diagram

Apollo vs. Artemis

Before diving into what makes Artemis different, it helps to see the programs side by side:

Dimension	Apollo (1961-1972)	Artemis (2017-present)
Primary Goal	Crewed lunar landing	Sustainable, international presence on and around the Moon
Mission Duration	8-12 days	Multi-week missions with eventual long-duration stays
Risk Posture	Accept high risk to meet a geopolitical deadline	Manage risk methodically; build capability iteratively
International Character	Predominantly American with limited cooperation	60-nation coalition through Artemis Accords
Economic Model	Government-funded and operated	Hybrid: government rockets with commercial landers, suits, and services
Legacy	Demonstrated humans could land on and return from the Moon; established technological prestige	Create infrastructure (SLS, Orion, Gateway, landers) and frameworks for sustained deep-space presence

The temptation to draw direct parallels between Apollo and Artemis is understandable—both involve flying humans to the Moon. But the similarities mostly end there. Apollo was born of Cold War urgency and executed with a singular national focus that accepted high risk for speed. Once the goal was achieved, funding evaporated and the hardware was mothballed. The Saturn V rocket, one of humanity's greatest engineering achievements, made its final flight in 1973 and was never built again.

Artemis represents something fundamentally different: a long-term strategy to build reusable systems, forge international partnerships, and establish normative frameworks for how nations will share the lunar frontier. The vehicles are more complex, the risk posture more conservative, and the ambitions extend far beyond symbolic footprints. Where Apollo asked "Can we get there?", Artemis asks "Can we stay?"

The Artemis II Crew

Before exploring what Artemis aims to accomplish, it's essential to understand who will be accomplishing it. Four astronauts will fly Artemis II, and their selection represents more than technical qualification—it signals that the future of exploration belongs to everyone willing to do the work.

Reid Weismann
Commander

Victor Glover
Pilot

Christina Koch
Mission Specialist

Jeremy Hansen
Mission Specialist

(Photos: Courtesy of NASA)

Cmdr Reid Wiseman inside the International Space Station - 2014.
(Photo courtesy of NASA)

Reid Wiseman
Commander

Reid Wiseman commands the mission, bringing experience from six months aboard the International Space Station and years as NASA's chief astronaut. A U.S. Navy aviator and test pilot, he flew forty-seven combat missions and holds a master's degree in systems engineering. His leadership style emphasizes collaboration and calm under pressure—qualities essential for a mission where the crew will be farther from Earth than any humans in more than fifty years. Wiseman's job is to bring everyone home safely while accomplishing as many objectives as possible, making the final calls when situations go off-script.

Victor Glover at Johnson Space Center - 2022.
(Photo courtesy of NASA)

Victor Glover
Pilot

Victor Glover serves as pilot. He flew on SpaceX's first operational Crew Dragon mission and spent 168 days on the space station conducting experiments and spacewalks. Artemis II will make him the first Black astronaut to travel beyond low Earth orbit. For young people of color watching from Earth, Glover's presence sends a powerful message: this frontier is yours too. During the mission, he'll pilot Orion manually during the proximity operations demonstration—the most demanding test of the spacecraft's handling characteristics, evaluating how the vehicle responds to manual commands and providing data that will inform training for future crews.

Specialist Christina Koch aboard the ISS - 2019.
(Photo courtesy of NASA, CC BY 2.0)

Christina Koch
Mission Specialist

Christina Koch will oversee the mission's scientific experiments and assist with navigation. She holds the record for the longest single spaceflight by a woman—328 days—and participated in the first all-female spacewalk. Koch's journey to NASA included years working in remote research stations and designing instruments for space missions. She will be the first woman to fly to the Moon, a milestone that matters because representation shapes who sees themselves in these roles. In Orion's cramped 9 cubic meters of habitable space, Koch will manage life support systems that must function flawlessly for ten days, monitoring oxygen levels, carbon dioxide scrubbing, and water recycling in an environment where failure means disaster.

Jeremy Hansen - 2022.
(Photo credit Nattanon Dungsunenarn/Spaceth, CC BY 2.0)

Jeremy Hansen
Mission Specialist

Jeremy Hansen rounds out the crew as the first Canadian to leave Earth orbit. A Royal Canadian Air Force colonel and CF-18 fighter pilot, Hansen has trained NASA's astronaut classes and served as a capsule communicator for space station operations. His inclusion reflects Canada's contributions to Artemis—robotic arms for Gateway, partnership on rovers, and decades of collaboration on human spaceflight. When Hansen looks back at Earth from lunar distance, he'll carry the hopes of an entire nation that has invested in this vision of cooperative exploration.

Together, these four astronauts represent a departure from the Apollo era's narrow demographic. They come from different backgrounds, different countries, and different life experiences. They've trained together for years, building the trust and communication rhythms required to operate a spacecraft where split-second decisions can mean the difference between success and catastrophe. Their mission won't land on the Moon, but it will prove that a diverse crew can function seamlessly in deep space—a prerequisite for the even more ambitious missions that will follow.

Building the Pipeline

NASA frames Artemis as part of a Moon-to-Mars architecture—a blueprint for sustained human presence beyond Earth. The core idea is straightforward: use the Moon as a proving ground for the technologies, operations, and international cooperation that will eventually be required for Mars. The Moon is close enough that resupply missions can reach it in days and communication delays measure in seconds rather than minutes. It's an ideal testbed before committing to the eight-month journey to Mars, where no rescue mission can arrive in time if something goes wrong.

This strategy involves several interconnected objectives. First, NASA must build a reliable transportation pipeline—rockets, capsules, and landers that can ferry crews and cargo between Earth and the lunar surface on a regular cadence. The Space Launch System rocket and Orion capsule form the backbone of this pipeline, with commercial Human Landing Systems from companies like SpaceX and Blue Origin providing the final leg to the surface. The Gateway space station, a small outpost in lunar orbit, will serve as a staging point and research platform. Each mission is designed to refine hardware and procedures so that failures are resolved before they threaten long-duration exploration.

Artemis II plays a critical role in this pipeline validation. Commander Wiseman and the crew must demonstrate that Orion's systems—life support, navigation, communications, propulsion—perform reliably with humans aboard for an extended deep-space mission. The proximity operations demonstration, where pilot Glover will manually fly Orion around the spent upper stage, tests whether astronauts can assume control if automation fails during future docking operations with Gateway or lunar landers. Every system checkout, every manual procedure, every anomaly the crew encounters and resolves provides data that makes subsequent missions safer and more capable.

Second, Artemis aims to develop robust surface infrastructure. Unlike Apollo's three-day visits, the program envisions missions lasting weeks to months. That requires habitats that can support crews through the two-week lunar night, pressurized rovers that allow astronauts to live and work inside them for extended periods, reliable power systems including nuclear fission reactors, and techniques for extracting water and oxygen from lunar ice—a practice called in-situ resource utilization, or ISRU. These capabilities will eventually enable manufacturing and mining at the south pole, transforming the Moon from a destination you visit into a place where you can actually live and work.

Third, the program deliberately fosters international cooperation. More than sixty nations have joined the Artemis Accords, a set of principles for peaceful exploration that emphasize transparency, interoperability, emergency assistance, and preservation of lunar heritage sites. By inviting a broad coalition to sign on and contribute hardware—Europe provides

Orion's service module, Japan is building life-support systems and a pressurized rover, Canada supplies robotic arms—NASA hopes to create diplomatic ties and shared ownership in the endeavor. Jeremy Hansen's presence on Artemis II embodies this partnership model, demonstrating that deep-space exploration is genuinely international rather than American-dominated. This approach spreads costs and expertise while establishing norms before the Moon becomes crowded and contested.

Finally, every aspect of Artemis is designed with Mars in mind. Regenerative life-support systems that recycle water and air, advanced communications networks, autonomous navigation software, and crew health studies all serve as testbeds for future Mars expeditions where resupply is impossible and communication delays stretch to twenty minutes round-trip. The Gateway's orbit—a gravitationally stable path that swoops high above the Moon—is a prototype for the deep-space habitats that will eventually ferry crews to the Red Planet. Christina Koch's extensive experience with long-duration spaceflight—328 days aboard the International Space Station—provides essential perspective on maintaining crew health and performance during extended missions. By the early 2030s, NASA intends to apply these lessons to send the first humans to Mars.

The Artemis flag at Kennedy Space Center in Florida on Jan. 6, 2023
(Photo courtesy of NASA/Glenn Benson)

Competition v Cooperation

During the Apollo era, the Soviet Union was the singular competitor. Today the landscape is crowded with nations and companies racing to stake claims on the lunar frontier. The most immediate competitor is China, which has declared plans to land taikonauts near the lunar south pole by 2030. That timeline matters because the south pole is unique: certain crater rims receive near-constant sunlight for power, while nearby permanently shadowed regions may harbor water ice. Access to these resources could determine which nations can refuel spacecraft, manufacture propellant on-site, and support sustained human presence. If China arrives first, it may set precedents for resource extraction and traffic management that other nations will find difficult to challenge.

India has also demonstrated ambitions, landing a rover near the south pole in 2023. Russia, despite economic constraints, continues to pursue lunar missions and has partnered with China on plans for a joint research station. Private companies add another layer of complexity: SpaceX is developing Starship as a fully reusable heavy-lift vehicle that could dramatically reduce launch costs; Blue Origin is building lunar landers; and dozens of startups are working on everything from lunar communications networks to mining equipment. The Moon, once the province of two superpowers, is becoming a crowded frontier.

The United States government positions Artemis as a

demonstration of leadership in this multipolar environment. By inviting a broad coalition to participate and sign the Artemis Accords, NASA aims to shape how the lunar commons will be governed before competition escalates into conflict. At stake is more than national prestige. Control of—or equitable access to—ice deposits could enable a cislunar economy where spacecraft refuel in lunar orbit, habitats generate their own oxygen, and mining operations extract metals for construction. The nation or coalition that establishes this infrastructure first may define the rules of engagement for decades.

At the same time, cooperation offers pathways that pure competition does not. By pooling resources and expertise through the Artemis Accords, participating nations can achieve collectively what none could manage alone. Europe's contribution of the Orion service module, for instance, reduced NASA's development costs and secured European astronauts seats on future missions. Japan's pressurized rover will expand exploration range while offering Japanese scientists access to lunar samples. Canada's robotic systems for Gateway buy Canadian astronauts a place in the crew rotation. These exchanges weave a web of interdependence that makes unilateral action more difficult and builds constituencies within each nation that support continued collaboration.

The question is whether this coalition model can hold. China and Russia have not signed the Artemis Accords, viewing them as a Western-led framework that doesn't adequately reflect their interests. They are pursuing their own International Lunar Research Station concept and building partnerships with nations that may see advantages in a multipolar approach to lunar governance. Whether the Moon becomes a domain of peaceful cooperation or strategic competition remains an open question—but the answer will be shaped in large part by the success or failure of the Artemis missions now underway.

The Artemis Accords

The Artemis Accords were established in 2020 and have since grown to include more than sixty signatory nations—roughly one-third of all United Nations member states. They represent a non-binding set of principles that supplement the 1967 Outer Space Treaty, which prohibits national appropriation of celestial bodies but left many questions unanswered about resource use, traffic management, and heritage preservation. The Accords attempt to fill those gaps by articulating norms for responsible behavior in space.

The first principle is peaceful purposes. Signatories commit to using space exclusively for peaceful activities and to conduct exploration in accordance with international law. This doesn't prohibit military organizations from participating—many astronauts are military officers, and launch facilities are often operated by defense agencies—but it establishes that aggression and weaponization have no place in lunar exploration.

Transparency is the second pillar. Nations agree to publish their policies, share scientific data openly, and communicate their plans to avoid misunderstandings. In practice, this means that when a nation targets a landing site or deploys a rover, it informs other signatories so missions don't inadvertently interfere with one another. The principle builds trust and reduces the risk of accidental collisions or disputes over who arrived first.

Interoperability follows naturally from transparency.

Signatories commit to using common standards wherever possible—docking ports, communications protocols, power connectors—so that hardware from different nations can work together. This isn't just about convenience; it's about safety. If an astronaut from one nation needs emergency assistance, equipment from another nation's mission should be able to provide it without requiring adapters or workarounds.

Emergency assistance itself is a fourth principle. Nations pledge to render aid to astronauts in distress, regardless of nationality. This echoes maritime law's obligation to rescue those in peril at sea and recognizes that space remains an unforgiving environment where every crew is vulnerable.

Registration and data sharing ensure that space objects are tracked and scientific findings are made public. Nations must register spacecraft and infrastructure with international authorities, and they agree to release data in formats that other researchers can use. This openness accelerates scientific progress and ensures that taxpayer-funded exploration benefits humanity broadly.

Heritage preservation addresses a concern unique to space: the risk of damaging historically significant sites. The Apollo landing areas, for instance, contain footprints, equipment, and flags that represent humanity's first steps on another world. The Accords commit signatories to respecting these sites and avoiding actions that could disturb them. As lunar traffic increases, this principle may evolve into designated heritage zones with specific access restrictions.

Finally, the Accords address orbital debris and deconfliction. Signatories agree to minimize space debris—spent rocket stages, broken equipment, fragments from collisions—and to coordinate activities through "safety zones" that prevent harmful interference. If one nation is conducting delicate operations at a particular crater, others agree to avoid flying overhead or landing nearby until the work is complete.

Nineteen countries joined the Accords in 2024 alone, reflecting a surge of interest as Artemis missions approach and the prospect of lunar resource utilization becomes tangible. Europe accounts for twenty-nine signatories; the Americas, Africa, Asia, and Oceania are all represented. The coalition includes spacefaring powers like Japan and the United Kingdom alongside nations like Latvia, Rwanda, and the Philippines that are just beginning to develop space programs. For these smaller countries, signing the Accords offers a seat at the table and a voice in shaping norms that will govern the cislunar economy for decades to come.

Critics note that the Accords are non-binding and lack enforcement mechanisms. If a signatory violates the principles—by, say, refusing to share data or failing to deconflict operations—there are no penalties beyond diplomatic pressure and potential exclusion from future collaborations. Skeptics also point out that the two nations most likely to challenge Western dominance in space, China and Russia, have not signed and show no intention of doing so. Whether the Accords become the foundation for a truly global governance framework or remain a coalition of the willing depends on how inclusive and adaptable they prove to be as lunar activity intensifies.

NASA's SLS rocket carrying the Orion spacecraft launches on the Artemis I flight test, Wednesday, Nov. 16, 2022
(Photo courtesy of NASA/Keegan Barber)

CHAPTER TWO

Artemis I

Before trusting the Space Launch System and Orion with human lives, NASA conducted an uncrewed flight test. Artemis I launched in November 2022 and sent an Orion capsule on a 25-day journey around the Moon and back. The mission aimed to verify that the integrated system—the massive rocket, the European-built service module, and Orion's avionics and heat shield—could perform as designed under the extreme conditions of deep space.

The results exceeded expectations in almost every measurable way. The rocket delivered Orion to its target orbit with pinpoint accuracy. The spacecraft completed 161 planned test objectives and added twenty more during flight as controllers seized opportunities to push systems beyond their original parameters. Dynamic events that engineers had worried about for years—tower jettison, booster separation, parachute deployment—executed without issue. Orion splashed down within a few miles of its target zone in the Pacific Ocean.

The European Service Module generated about twenty percent more electrical power than predicted while consuming roughly twenty-five percent less propellant. This margin meant future missions could carry additional payload or extend their duration. Navigation systems performed flawlessly, and the Deep Space Network maintained communications even when Orion swung behind the Moon's far side. The spacecraft's guidance computer handled the complex orbital mechanics of a distant retrograde orbit—a high, stable

Orion's heat shield following Artemis I
(Photo courtesy of NASA)

path that carried Orion nearly 65,000 kilometers beyond the Moon, farther from Earth than any human-rated spacecraft had ever traveled.

The mission also demonstrated NASA's ability to adapt in real time. When engineers identified opportunities to test additional modes and gather extra data, ground controllers uplinked new commands. They tested optical communications equipment at higher data rates than originally planned, deliberately induced stress on thermal control systems to see how they responded, and practiced contingency procedures for scenarios that future crews might encounter. This aggressive testing philosophy—fly the mission you planned, but also fly the mission you discover—proved that Artemis I was more than a checkout flight. It was an active learning laboratory.

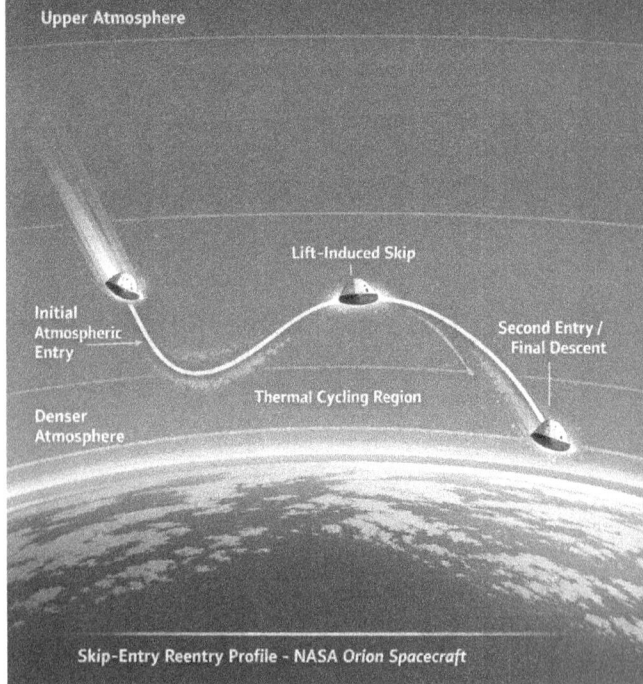

Skip-Entry Reentry Profile

Upper Atmosphere

Lift-Induced Skip

Initial
Atmospheric
Entry

Second Entry /
Final Descent

Thermal Cycling Region

Denser
Atmosphere

Skip-Entry Reentry Profile - NASA Orion Spacecraft

Skip-Entry Reentry Profile
(Infographic by MS Designer)

The Heat Shield Crisis

Then came the discovery that would force NASA into one of the most consequential engineering decisions of the Artemis program.

As Orion slammed into Earth's atmosphere at nearly 25,000 miles per hour, sensors and cameras recorded something that no ground test had predicted: chunks of the heat shield's ablative material flaked off asymmetrically, leaving irregular divots and exposed areas across the shield's surface. The heat shield, covered in a material called Avcoat, is designed to char and erode during reentry—that charring carries heat away from the spacecraft through ablation. But the char should ablate evenly, creating a uniform protective layer. Instead, investigators found material breaking away in unpredictable patterns.

A post-flight investigation traced the problem to gases trapped within the Avcoat's honeycomb structure. The heat shield is built from thousands of small cells, each filled with Avcoat material. During the skip-entry profile that Artemis I flew—where the capsule dipped into the atmosphere, then briefly skipped back into space before reentering fully—heating rates varied dramatically. When temperatures dropped during the skip phase, gases generated by ablation couldn't escape. As the capsule descended again, those trapped gases built up pressure under the char layer, eventually causing chunks to spall off in irregular patterns.

Ground tests had not captured this phenomenon

because test facilities produce different heating profiles than actual flight. Arc-jet facilities, which blast spacecraft materials with superheated plasma, tend to maintain more constant heating. The gases vent continuously rather than getting trapped. It took a real mission, flying a real skip-entry trajectory, to reveal the problem.

The immediate question was whether the char loss threatened crew safety. Temperatures inside the cabin had remained comfortable throughout reentry, and the underlying heat shield material stayed intact. The system worked—but it didn't work as predicted. More critically, the unpredictable nature of the char loss meant NASA couldn't confidently model where material might flake away on future flights or how much margin existed before the spalling compromised structural integrity.

NASA faced a crisis that would define the Artemis II mission.

The Impossible Choice

The discovery left NASA leadership confronting a decision that would shape the entire program's trajectory: commit to an 18-month delay to completely replace Orion's heat shield, or proceed with a fundamental change to how the crew would return to Earth.

Neither option was acceptable. Both were necessary to consider.

Option A: Replace the heat shield entirely. Each heat shield takes many months to manufacture—technicians hand-fill each honeycomb cell, cure the material under precise temperature and pressure conditions, and machine the final surface to exacting tolerances. Integrating a new shield would require disassembling much of the spacecraft, testing the connections, and revalidating systems that had already been certified. The delay would ripple through the entire program: Artemis III's lunar landing would slip, international partners' contributions would sit idle, and the competitive race with China—which aims to land taikonauts by 2030—would shift decisively in Beijing's favor. Eighteen months might sound manageable, but in the budget cycles of Congress and the attention span of political support, it represented existential risk.

Option B: Modify the reentry trajectory. Engineers proposed a lofted entry profile—a single-dip reentry that would eliminate the thermal cycling responsible for gas trapping. Think of it this way: the skip entry used on Artemis I was like skipping a stone across water to extend how far it travels, creating multiple heating and cooling phases. The lofted entry is more like dropping the stone straight in—one continuous descent with no intermediate plateau where gases could accumulate and build pressure.

But the lofted entry extracted a price. The crew would experience higher peak temperatures—around 2,760 degrees Celsius—as the capsule followed a steeper, faster descent through the atmosphere. The G-forces would intensify during the more aggressive trajectory. And most significantly, this profile had never been tested with actual hardware. Ground simulations suggested it would work. Models indicated the heat shield could handle the increased thermal load. But certainty wouldn't come until an actual spacecraft flew the profile.

NASA was choosing between two deeply undesirable options: proceed with a trajectory that had never been flown by this vehicle, accepting known, quantifiable risks of higher temperatures and G-loads, or fly the same skip-entry profile with an unknown risk of unpredictable char loss that could cascade unpredictably.

The decision revealed NASA's risk management philosophy developed over decades of human spaceflight, refined by the Challenger and Columbia disasters, and validated through countless engineering reviews: when confronted with uncertainty, choose the well-understood danger over the unpredictable one.

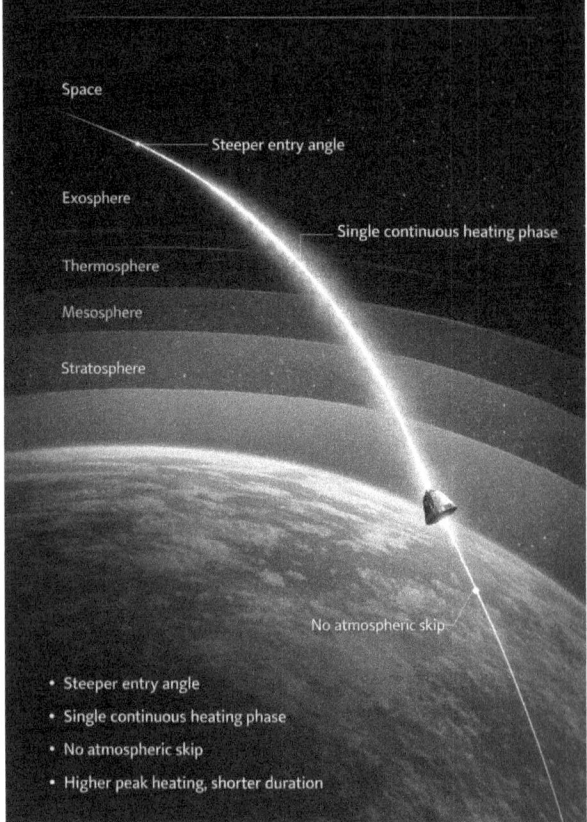

Modified Lofted-Entry Atmospheric Reentry

Space

Steeper entry angle

Exosphere

Single continuous heating phase

Thermosphere

Mesosphere

Stratosphere

No atmospheric skip

- Steeper entry angle
- Single continuous heating phase
- No atmospheric skip
- Higher peak heating, shorter duration

Modified Lofted-Entry Reentry Profile
(Infographic by MS Designer)

Known Risk v Unknown Risk

The logic was rigorous but uncomfortable. The skip entry that Artemis I flew had produced asymmetric char loss—unpredictable in location and magnitude. Engineers couldn't confidently model where material would flake away on a crewed flight or whether the pattern might worsen. The gas trapping mechanism was understood after the fact, but predicting its behavior on subsequent flights remained uncertain. That uncertainty—the inability to bound the risk—made the skip entry unacceptable for carrying crew.

The lofted entry, by contrast, offered risks that could be quantified and managed. Yes, peak heating would reach 2,760 degrees Celsius. Yes, the crew would experience higher G-forces during the steeper descent. But engineers could model these conditions precisely. They knew how Avcoat performed at those temperatures. They understood the structural loads. They could calculate margins and verify that the spacecraft would protect the crew. The risks were higher in absolute terms, but they were known risks that could be tested, analyzed, and accepted based on data rather than hope.

Moreover, the lofted entry addressed the root cause. By eliminating the skip phase, the profile prevented the thermal cycling that trapped gases in the first place. The heating would be more intense but more uniform. Gases generated during ablation could vent continuously as the char layer formed, preventing the pressure buildup that caused spalling. The physics worked in

NASA's favor—trading one severe but brief thermal pulse for the elimination of an unpredictable failure mode.

The decision to proceed with the lofted entry was made in mid-2024 after exhaustive analysis. Engineers refined Avcoat manufacturing processes for future missions, adding vent paths that allow gases to escape more easily and adjusting curing procedures to minimize gas generation. But for Artemis II, the combination of the existing heat shield and the modified trajectory provided adequate safety margins based on rigorous modeling.

What this meant for Commander Reid Wiseman, pilot Victor Glover, and mission specialists Christina Koch and Jeremy Hansen was stark: they would be flying a reentry profile that had never been tested with hardware. The confidence came from analysis, simulation, and decades of expertise—but not from prior flight experience with this specific trajectory. Artemis II thus transformed from a validation mission repeating Artemis I's success with crew aboard into something more demanding: the first test of an entirely new entry profile, one designed to overcome a failure mode that no amount of ground testing had predicted.

The crew understood the stakes. In media interviews before the mission, Wiseman acknowledged the calculated nature of the choice: "We're test pilots. Our job is to fly systems that haven't been flown before and validate that they work. The lofted entry is the right engineering decision, and we trust the analysis. But we also know we're proving it works in the only way that ultimately matters—by flying it."

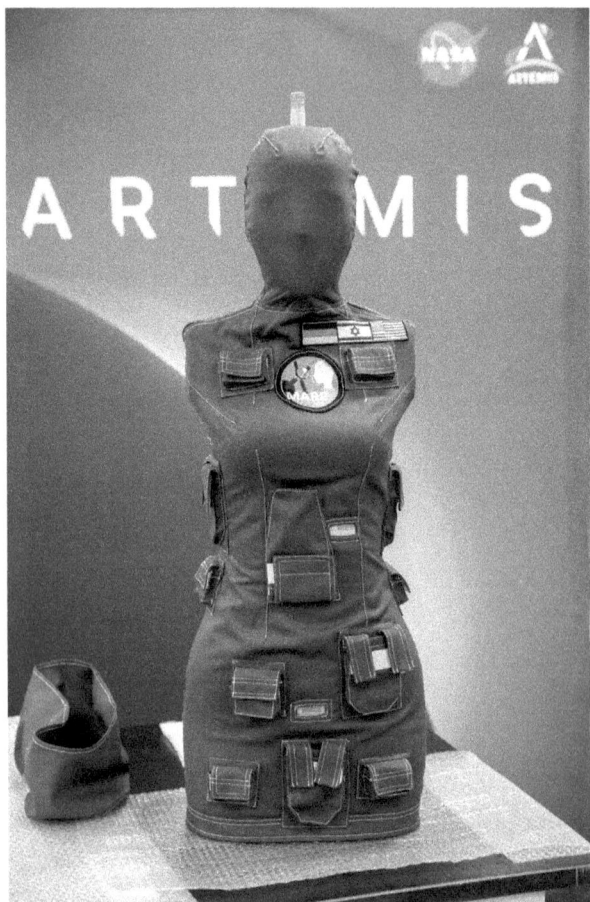

Helga, one of two identical phantom torsos, is shown inside the Space Station Processing Facility at Kennedy Space Center
(Photo courtesy of NASA/Kim Shiflett)

Additional Lessons Learned

Beyond the heat shield, Artemis I revealed dozens of smaller discoveries that informed preparations for Artemis II. The mission carried ten CubeSat payloads—shoebox-sized satellites from universities, international partners, and private companies. These secondary missions aimed to demonstrate technologies ranging from solar sails to lunar landing techniques. Several failed due to battery degradation or deployment anomalies, but a few succeeded brilliantly.

The Italian ArgoMoon CubeSat captured images of the upper stage separation, providing valuable data on how debris behaves in deep space. NASA's BioSentinel carried yeast cells to study how DNA repair mechanisms respond to deep-space radiation, comparing flight samples with controls on Earth. These small missions underscored NASA's desire to democratize exploration—allowing universities and smaller nations to ride along on flagship launches and conduct meaningful science.

Inside Orion, biological experiments provided early data on deep-space hazards. The Matroshka Astro-Rad Radiation Experiment installed two female-bodied phantoms—artificial torsos modeled after human anatomy—filled with thousands of radiation sensors. One phantom wore an AstroRad vest, a protective garment designed to shield astronauts from solar particle events. By comparing sensor readings between the shielded and unshielded phantom, researchers measured how much radiation dose could be reduced by wearable protection. The data informed decisions about suit design, habitat shielding, and crew limits for long-duration missions.

The flight also validated communications and navigation techniques essential for crewed missions. Engineers learned how to manage communications blackouts when Orion passes behind the Moon, how to uplink commands through the Deep Space Network with minimal latency, and how autonomous systems behave when they can't check in with ground controllers. These operational lessons—unglamorous but critical—built confidence that crews could function effectively even when separated from Earth by distance and occasional silence.

Preparing Artemis II

The progression from Artemis I to Artemis II illustrates NASA's methodical strategy. By flying uncrewed first, engineers identified and addressed issues that would have been unacceptable on a crewed flight: the heat-shield gas venting, electromagnetic interference with certain sensors, minor thruster anomalies. The data collected on radiation exposure, micro-meteorite impacts, and long-duration avionics performance provides empirical evidence to refine models and improve designs.

But Artemis II is emphatically not simply Artemis I with people aboard. The presence of a crew changes almost every aspect of the mission, and the switch to a lofted entry profile means the crew will be validating a trajectory that has never been flown by this spacecraft.

Life-support systems must operate continuously for ten days, recycling water and scrubbing carbon dioxide from the cabin air. The uncrewed flight used minimal consumables; the crewed mission will stress the Environmental Control and Life Support System to its limits. Engineers have added crew displays, audio communications equipment, and an exercise harness to maintain the astronauts' physical conditioning. A docking adapter and sensors allow the crew to practice rendezvous procedures during the proximity operations demonstration, preparing them for future docking with Gateway or a lunar lander.

The reentry profile shift from skip entry to lofted entry

represents the most significant operational change. Mission planners have re-targeted the splashdown zone to ensure the capsule lands within reach of recovery forces. Weather criteria and launch windows have been adjusted accordingly. The heat shield itself carries more instrumentation—additional temperature and pressure sensors will capture fine-grained data during reentry, allowing engineers to validate their models and confirm that the lofted entry performs as predicted.

Christina Koch, who will monitor environmental systems throughout the mission, has trained extensively on the life-support interfaces. She'll watch carbon dioxide levels, oxygen partial pressure, humidity, and thermal control performance, comparing actual readings against predicted values. Victor Glover has spent hundreds of hours in simulators practicing the manual proximity operations demonstration, learning how Orion responds to hand controller inputs and building the muscle memory required to fly the spacecraft smoothly around the spent upper stage. Reid Wiseman has rehearsed emergency procedures for scenarios ranging from cabin depressurization to main engine failures, ensuring the crew can respond effectively even during the communications delays and blackouts that characterize deep-space operations.

Ground operations have evolved as well. The mobile launch tower has been reinforced after the damage sustained during Artemis I's liftoff. An emergency egress system—baskets riding a zip-line from the crew access arm to a hardened bunker—has been installed and tested. The crew participated in a full Countdown Demonstration Test in December 2025, donning spacesuits, boarding Orion, and rehearsing the final hours before launch. These dry runs reduce the risk of first-time errors when the countdown is real.

Feature	Artemis I (2022)	Artemis II (2026)
Crew	None	Four astronauts
Mission Duration	25.5 days	~10 days
Entry Profile	Skip entry (tested)	Lofted entry (validation flight)
Entry Profile Risk	Unpredictable char loss discovered	Higher peak temps (2,760°C) and G-loads, but modeled and understood
Life Support	Minimal systems	Full environmental control with recycling
Proximity Operations	None	Manual piloting around spent upper stage
Primary Objectives	Validate SLS, Orion, ESM	Test crew systems, validate lofted entry, manual control, deep-space operations
Heat Shield	Baseline Avcoat; skip entry	Modified vent paths; lofted entry (first flight)
Launch Tower	Baseline; damage discovered	Reinforced with egress system

Opposite: NASA's Orion spacecraft splashes down in the Pacific Ocean aft
a 25.5 day mission to the Moon.
(Photo courtesy of NASA/Josh Valcarcel)

Why Test Flights Matter

The history of human spaceflight shows that systems rarely work perfectly on the first attempt. NASA Administrator Bill Nelson has emphasized that Artemis is technically challenging and that the agency must "get Artemis II right" before moving to lunar landings. By flying uncrewed first, NASA verified that the Space Launch System could deliver Orion to the Moon with precision, that the European Service Module could generate more power and use less propellant than expected, and that critical events like tower jettison and parachute deployment worked as designed.

The uncrewed mission also revealed what no amount of ground testing could replicate: the heat shield's behavior during a skip-entry profile in actual flight conditions. Without that discovery, a crewed flight would have carried unacceptable risk. The decision to switch to a lofted entry, while introducing its own challenges, demonstrated that NASA could identify problems, analyze alternatives under pressure, and make calculated choices that prioritized crew safety even when those choices introduced schedule and political risks.

Artemis I demonstrated something else equally important: that NASA can build, integrate, and fly a complex deep-space system in an era when such undertakings are often dismissed as too expensive or too difficult. The mission showed that international partnerships—Europe's service module, contributions from dozens of nations—can function seamlessly. It proved that fifty years after Apollo, the engineering knowledge to leave Earth orbit and return safely still exists within NASA and its contractors, even if much of that knowledge had to be rebuilt from scratch.

The uncrewed mission also revealed how much remains to be learned. Deep space is unforgiving. Radiation levels fluctuate unpredictably. Micro-meteorites pepper spacecraft at velocities that ground tests can only approximate. Communication delays and autonomous operations create scenarios where crews must make decisions without waiting for instructions from Earth. Every test flight reduces uncertainty and builds confidence, but perfection remains elusive. The goal is not to eliminate all risk—that's impossible—but to understand risk well enough to manage it.

As Artemis II approaches, the lessons from the first flight have been absorbed into hardware, software, and procedures. Engineers know what the heat shield will endure during the lofted entry, even though the profile has never been flown. Controllers understand how to manage communications blackouts. Recovery teams have practiced extracting crews from the Pacific. The mission carries forward everything learned from Artemis I while adding the complexity and unpredictability of human operators and the validation of an entry profile born from engineering necessity.

If it succeeds, NASA will certify the system for the lunar landings that follow. If challenges arise, the program will adapt again—learning, refining, and building toward the sustained presence that justifies the entire endeavor. But the crew strapping into Orion will know they're not repeating what's been done before. They're proving that calculated risks, when understood and managed properly, can be accepted—and that the path to sustained exploration requires the courage to fly profiles that exist only in simulations until someone demonstrates they work in reality.

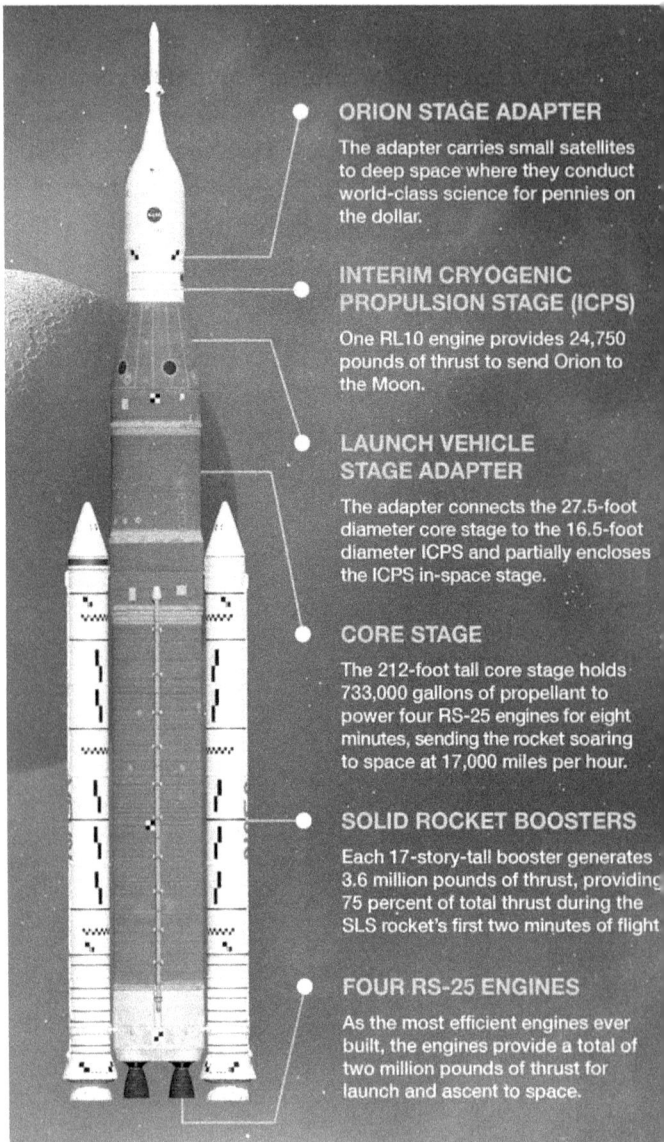

ORION STAGE ADAPTER

The adapter carries small satellites to deep space where they conduct world-class science for pennies on the dollar.

INTERIM CRYOGENIC PROPULSION STAGE (ICPS)

One RL10 engine provides 24,750 pounds of thrust to send Orion to the Moon.

LAUNCH VEHICLE STAGE ADAPTER

The adapter connects the 27.5-foot diameter core stage to the 16.5-foot diameter ICPS and partially encloses the ICPS in-space stage.

CORE STAGE

The 212-foot tall core stage holds 733,000 gallons of propellant to power four RS-25 engines for eight minutes, sending the rocket soaring to space at 17,000 miles per hour.

SOLID ROCKET BOOSTERS

Each 17-story-tall booster generates 3.6 million pounds of thrust, providing 75 percent of total thrust during the SLS rocket's first two minutes of flight

FOUR RS-25 ENGINES

As the most efficient engines ever built, the engines provide a total of two million pounds of thrust for launch and ascent to space.

Opposite: A diagram of NASA's SLS rocket carrying the Orion spacecraft.
(Graphic courtesy of NASA)

CHAPTER THREE

The Stack: SLS + Orion

Four astronauts will ride this machine into deep space. Reid Wiseman, Victor Glover, Christina Koch, and Jeremy Hansen will strap into seats atop a tower of propellant and metal that stands as tall as a 32-story building and generates more thrust at liftoff than any rocket America has built since Saturn V. Their survival depends on every component of this integrated stack working flawlessly—from the solid rocket boosters that hurl them away from Earth to the European Service Module that will sustain them for ten days to the heat shield that will protect them during the inferno of reentry.

The Artemis II launch vehicle is not a single rocket but a carefully integrated stack of subsystems working in concert. At the base sits the Space Launch System—the most powerful rocket NASA has built since Saturn V. Atop that rides the Orion crew capsule, humanity's first deep-space spacecraft since Apollo. Behind Orion, providing the muscle to actually reach the Moon, is the European Service Module built by Airbus for the European Space Agency. Understanding how these pieces fit together illuminates both the mission's technical challenges and the choices NASA made about what to build, what to reuse, and what to buy from partners.

SPACE LAUNCH SYSTEM

ARTEMIS II:
FOUR RS-25 ENGINES

UPGRADED & READY FOR A BOLD NEW MISSION

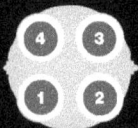

FOUR RS-25 ENGINES are arranged at the bottom of the **SLS** (Space Launch System) rocket. These engines are specially designed to accommodate the different operational and environmental conditions to power the Artemis missions to deep space.

SLS CORE STAGE

ENGINE SECTION **PROPELLANT TANKS**

WHAT'S DIFFERENT?

THREE RS-25 ENGINES were required for each space shuttle mission to low Earth orbit. The engines operated for 8.5 minutes and created a combined **1.4 MILLION POUNDS OF THRUST.**

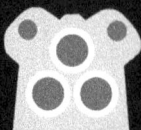

FOR SLS, the RS-25 power level was increased to give the NASA Moon rocket's **FOUR-ENGINE CLUSTER** more than
2 MILLION POUNDS OF THRUST.

www.nasa.gov/SLS   #ARTEMIS

Infographic featuring NASA's RS-45 Engines.
(Graphic courtesy of NASA)

Space Launch System (SLS)

The Space Launch System stands 98 meters tall—roughly the height of a 32-story building—and generates 8.8 million pounds of thrust at liftoff, about fifteen percent more than Saturn V managed. That thrust comes from two sources: a massive core stage powered by four RS-25 engines burning liquid hydrogen and liquid oxygen, and two enormous solid rocket boosters strapped to the core's sides. Every pound of that thrust exists for one purpose: to accelerate Orion and its crew fast enough to escape Earth's gravitational grip and send them toward the Moon.

The crew will feel that power viscerally. Wiseman and Glover, seated in the front row, will watch displays showing engine performance and trajectory as the vehicle climbs. Koch and Hansen, in the center seats behind them, will monitor systems that become critical once the rocket reaches orbit. For two minutes, the solid boosters will create vibration that shakes the entire stack—astronauts describe it as riding inside a paint mixer. When the boosters separate at 150,000 feet, the ride smooths dramatically, but acceleration continues building as the RS-25 engines push toward orbit.

NASA made a deliberate choice to base the rocket on Space Shuttle hardware rather than design something entirely new. The RS-25 engines are refurbished units that once powered the shuttle, upgraded with modern controllers and insulation. The five-segment solid rocket boosters are direct descendants of the shuttle's four-segment boosters, stretched to provide more thrust and burn time. Even the manufacturing facilities—the Michoud Assembly Facility in Louisiana where the core stage is built, the Kennedy Space Center integration buildings—are repurposed from the shuttle era.

Core Stage Infographic
(Graphic courtesy of NASA)

This evolutionary approach reduced development risk but locked NASA into a costly architecture. Each RS-25 engine requires extensive refurbishment between flights, and the solid boosters must be manufactured fresh for each launch since they cannot be reused in the same way SpaceX reuses its Falcon 9 first stages. Critics argue that this heritage hardware costs more per launch than commercial alternatives could deliver. Proponents counter that it offers proven reliability essential for crewed missions and provides capabilities—particularly payload mass to the Moon—that commercial vehicles have yet to demonstrate.

The rocket's upper stage for Artemis II is the Interim Cryogenic Propulsion Stage, built by United Launch Alliance and powered by a single RL-10 engine. After the core stage and boosters finish their work, this upper stage takes over, pushing Orion from Earth orbit onto its trajectory toward the Moon. The stage must perform two critical burns: one to raise Orion's orbit to an elliptical path reaching 38,000 miles from Earth, and a second to circularize that orbit before the final translunar injection. Victor Glover will monitor these burns closely, watching propellant flow rates and trajectory updates, ready to take manual control if the automated sequence encounters problems.

The design philosophy reflects NASA's position after the shuttle program ended. The agency still employed thousands of engineers with deep expertise in cryogenic propulsion, large structures, and human-rating procedures. Manufacturing facilities sat ready to be repurposed. By leveraging that infrastructure rather than starting from scratch, NASA could fly sooner—though not cheaply—and maintain an industrial base that might otherwise have dispersed. Whether that tradeoff proves worthwhile depends partly on what follows: if Artemis leads to a sustained lunar program and eventually Mars missions, the investment makes sense. If the program stalls after a few flights, the cost per mission will look increasingly difficult to justify.

Orion and the ESM

Atop the rocket sits Orion, a capsule superficially similar to Apollo's command module but larger and far more capable. The crew module provides about nine cubic meters of habitable volume—roughly the interior space of a large minivan—where Wiseman, Glover, Koch, and Hansen will live and work for ten days. Christina Koch, drawing on her record 328 days aboard the International Space Station, will monitor life-support systems in this confined space, ensuring that oxygen levels, carbon dioxide scrubbing, humidity,

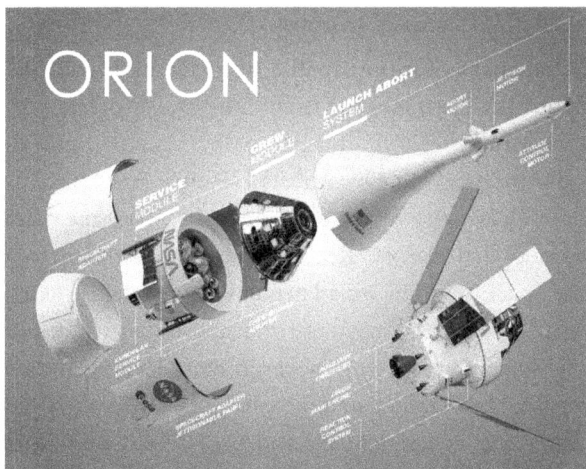

Orion Crew Module Diagram
(Graphic courtesy of NASA)

and temperature remain within safe parameters. Her experience with long-duration confinement and systems management makes her ideally suited for what amounts to running a miniature space station while simultaneously conducting experiments and assisting with navigation.

The crew module's primary structure is an aluminum-lithium pressure vessel covered by the Avcoat heat shield that will protect the crew during the fiery return through Earth's atmosphere. A side hatch allows entry and exit on the launch pad and during recovery. A launch abort tower sits on top, ready to yank the capsule away from the rocket if something goes catastrophically wrong during ascent. The capsule is designed for survival first, comfort second—but survival in deep space requires systems that Apollo never needed for its brief journeys.

What makes Orion capable of reaching the Moon is not the crew module itself but the European Service Module (ESM) attached below it. This cylindrical structure, built by Airbus Defense and Space, is the powerhouse of the spacecraft. It carries the main engine—an Aerojet Rocketdyne AJ10 derived from the Space Shuttle's orbital maneuvering system—along with 24 smaller thrusters for attitude control. Eight solar array wings, each seven meters long, generate about 11 kilowatts of electrical power. Tanks hold water, oxygen, nitrogen, and propellant. Radiators manage thermal control, keeping the spacecraft from overheating in direct sunlight or freezing in shadow.

Without the service module, Orion could not leave Earth orbit, navigate around the Moon, or sustain its crew. The module provides propulsion for the translunar injection burn that commits the spacecraft to lunar trajectory, for mid-course corrections along the way, and for the trajectory adjustments that fine-tune the return path. It supplies all electrical power once Orion separates from the rocket. It regulates temperature, distributes water, and manages the flow of oxygen to the crew module. When the mission ends, the service module is jettisoned before reentry—it burns up in the atmosphere while the crew module returns safely under parachutes.

The European Service Module is also the most visible expression of Artemis as an international endeavor. Jeremy Hansen's presence on the crew directly reflects this partnership—Canada's investment in Gateway robotics and contributions to life-support systems earned Canadian astronauts seats on Artemis missions. Europe's investment in building the hardware buys European astronauts similar opportunities and reduces NASA's development burden. The partnership has deep roots: European companies contributed to the International Space Station, and the European Space Agency has collaborated with NASA on planetary missions for decades. Extending that cooperation to Artemis ensures that when the program faces political challenges or budget pressures in the United States, European partners have a stake in seeing it continue.

For Hansen, the service module represents both opportunity and responsibility. He'll operate communications systems that depend on the module's power, assist with life-support monitoring that draws on the module's consumables, and serve as the liaison with

Canadian partners tracking the flight. When the crew performs the proximity operations demonstration, Hansen will help Glover and Wiseman coordinate the delicate dance of separating from the spent upper stage, backing away under automated control, and then switching to manual flight—all while the European Service Module provides the power and propulsion that makes those maneuvers possible.

Crew Systems

A capsule the size of a minivan becomes home for four astronauts over ten days. That requires systems that would seem mundane on Earth but become critical in the vacuum of space. Christina Koch, responsible for monitoring environmental controls throughout the mission, will spend significant time watching displays that show cabin air circulation, carbon dioxide levels, humidity, and temperature. The Environmental Control and Life Support System—often shortened to ECLSS or just "life support"—cir-

Inside the Orion Crew Module
(Photo courtesy of NASA)

culates cabin air through fans and filters, removes carbon dioxide using chemical scrubbers, and controls humidity to prevent condensation from forming on cold surfaces.

Water recycling turns condensation from the cabin

atmosphere, moisture from the crew's breath and sweat, and even urine into potable water through multi-stage filtration and distillation. The system produces drinking water for the crew and supplies moisture for food rehydration. Temperature and pressure sensors feed data to controllers that adjust heaters, pumps, and valves. If smoke or toxic fumes are detected, the crew can don portable breathing masks connected to emergency oxygen bottles. The system is designed with redundancy: backup pumps, alternate control pathways, emergency supplies.

But Artemis II marks the first long-duration test with a human crew in deep space. Engineers don't just want to see if the equipment functions—they want to understand how it responds to the unpredictable metabolic patterns of living humans in an environment far more hostile than low Earth orbit. Koch's 328 days on the space station taught her how life-support systems can behave unexpectedly as conditions change, how small anomalies can cascade if not caught early, and how crews must sometimes troubleshoot problems that ground controllers can't diagnose remotely. That experience becomes invaluable when the spacecraft is days from Earth and communication delays measure in seconds or minutes.

The crew will deliberately vary their activity levels during the mission. Wiseman and Glover will exercise vigorously during the proximity operations demonstration, generating heat, carbon dioxide, and humidity at high rates. Then they'll rest, allowing the system to recover. Koch and Hansen will monitor power consumption, oxygen partial pressure, and carbon dioxide levels throughout. These tests will validate performance models and identify any edge cases where the system struggles to maintain nominal conditions. Every data point Koch records becomes part of NASA's growing database on how life support behaves in deep space—information that will inform Gateway design and eventually Mars transit vehicles.

Living quarters are spartan. Four adjustable seats dominate the cabin during launch and reentry, positioned to maintain the crew's center of mass during acceleration. Once in space, the seats can be folded or reconfigured to create more room. Storage lockers line the walls, holding food, cameras, scientific instruments, personal items, and spare parts. A small galley includes a food warmer for heating thermostabilized meals—a significant improvement over Apollo's squeeze tubes and freeze-dried fare. A compact waste management compartment provides privacy for sanitation. An exercise harness allows crew members to perform resistance training, essential for mitigating bone and muscle loss even on a relatively short mission.

The cabin's design reflects lessons from decades of spaceflight. LED lighting can be adjusted to support circadian rhythms, helping astronauts sleep despite the absence of a natural day-night cycle. Windows provide views of Earth and the Moon—psychologically important during a confined mission—and allow the crew to perform visual navigation checks.

Touch-screen displays and physical controls provide redundancy: if a screen fails, essential commands can still be executed. Acoustic dampening reduces noise from pumps and fans, making communication easier and sleep more restorative.

In this confined space, the crew's diversity becomes an asset rather than a challenge. Wiseman's calm leadership, Koch's systems expertise, Glover's test pilot precision, and Hansen's international coordination experience create complementary skill sets that allow the crew to divide responsibilities effectively. Each has trained extensively on every system, but specialization emerges naturally: Koch gravitates toward life support and experiments, Glover focuses on flight controls and manual operations, Hansen manages communications and coordinates with international partners, and Wiseman orchestrates the overall mission flow while maintaining situational awareness across all systems.

Communications and Control

Navigating a spacecraft through deep space requires knowing precisely where you are, where you're going, and how to get there. Orion's guidance, navigation, and control system uses a suite of star trackers, inertial measurement units, gyroscopes, and accelerometers to determine position and orientation. Star trackers photograph the night sky and compare patterns against an onboard catalog, allowing the computer to calculate attitude even when GPS signals are unavailable—as they will be beyond Earth orbit. Inertial measurement units provide high-frequency data on rotational rates and linear acceleration, enabling the flight computer to maintain control between star-tracker updates.

Near Earth, GPS receivers estimate position and velocity. Farther out, the spacecraft relies on tracking by the Deep Space Network—NASA's global array of large radio antennas. By measuring the time it takes for signals to travel between Earth and Orion, and the Doppler shift in those signals, ground stations can determine the spacecraft's position to within a few meters. The crew can also perform manual navigation using sextant-like techniques, sighting on stars and Earth's horizon to cross-check the computer's calculations.

Control laws translate desired trajectory changes into commands for thrusters or the main engine. For translunar injection and mid-course corrections, the computer generates engine-firing sequences that satisfy trajectory constraints while keeping accelerations within human tolerance. The spacecraft's guidance system is designed with fault tolerance: if a sensor fails or readings disagree, software can isolate the faulty input and continue operating.

During the proximity operations demonstration, Victor Glover will test how well this automated system hands control to humans. The demonstration isn't just about flying around a spent rocket stage for the sake of it—it's about understanding how Orion handles under manual control, evaluating whether the fly-by-wire system feels natural to pilots, and collecting data that no amount of ground testing can replicate. Glover has spent hundreds of hours in simulators learning the spacecraft's response characteristics, but simulation can only approximate reality. When he takes the hand controllers three hours into the mission and commands Orion to separate from the upper stage, he'll be providing NASA with the first real-world data on how the spacecraft responds to human inputs in deep space.

If the vehicle responds sluggishly during certain maneuvers, or if thruster firings feel different than simulators predicted, Glover will report those discrepancies and engineers can refine the software before future crews attempt docking with Gateway or lunar landers. This validation becomes critical when you consider that Artemis III will require precise rendezvous and docking with SpaceX's Starship HLS in lunar orbit—a maneuver that must work flawlessly or the landing mission fails. Glover's test flight essentially certifies that astronauts can assume control when automation fails or when human judgment is required for complex operations.

Communication with Earth uses multiple frequencies and methods. S-band radio provides voice and basic telemetry. X-band supports higher data rates for engineering telemetry and stored science data. The Orion Artemis II Optical Communications System—a laser communications experiment—will downlink high-definition video at up to 260 megabits per second through a four-inch telescope mounted on the spacecraft. Gimballed mirrors track ground stations, maintaining the laser link even as the spacecraft rotates. This optical system is a pathfinder for future broadband communications from Mars, where data rates will determine how much science can be returned and how closely mission controllers can monitor spacecraft health.

As Orion travels to the Moon, communications transition from near-Earth ground stations to the Deep Space Network. Latencies stretch to several seconds round-trip, requiring the crew to time conversations and build patience. The spacecraft will experience communication blackouts when it passes behind the Moon, losing line-of-sight with Earth for up to an hour at a time. During these periods, Orion operates autonomously, following pre-programmed sequences and making decisions without ground input. The crew trains for these blackouts, understanding that they may need to troubleshoot problems or adjust plans without waiting for instructions from Houston. Wiseman's experience as chief astronaut, where he oversaw training and operations for the entire astronaut office, gives him perspective on how crews maintain situational awareness and make good decisions when separated from ground support.

Cost, Competition & Choices

The elephant in the room throughout any discussion of Artemis hardware is cost. NASA's Office of Inspector General estimated that each Space Launch System and Orion launch costs approximately $4.1 billion when development costs are amortized across early missions. Those numbers invite immediate comparison with commercial alternatives. SpaceX's Falcon Heavy, currently the most powerful operational rocket, launches for approximately $150 million in expendable configuration. The company's Starship, once operational, aims for dramatically lower costs through full reusability. Blue Origin is developing New Glenn with similar cost-reduction goals. China is developing the Long March 9. Why does NASA continue investing in expensive, partially expendable hardware when cheaper options may be on the horizon?

The answer has several layers. First, at the time Artemis II flies, those commercial alternatives have not yet demonstrated the capabilities required for crewed deep-space missions. Starship has not completed an orbital flight and recovery. New Glenn has not launched. The Long March 9 exists only on drawing boards. NASA argues that it needs a heavy-lift capability now—not five years from now—and that the Space Launch System provides that capability with hardware derived from systems that flew for three decades on the Space Shuttle. With China planning to land taikonauts by 2030, the geopolitical calculation becomes stark: spend $4.1 billion per launch to maintain leadership now, or spend less to fly later and cede the competitive window to Beijing.

Second, human-rating a launch vehicle imposes requirements that go beyond raw performance. The rocket must demonstrate extremely high reliability, provide abort modes throughout ascent, undergo exhaustive testing, and incorporate redundancy in critical systems. NASA's design margins for crewed vehicles are more conservative than those typically used by commercial operators willing to accept higher risk in exchange for rapid iteration. Whether that conservatism is justified or excessive depends on one's tolerance for risk and perspective on the value of human life in exploration. But when you're launching Reid Wiseman, Victor Glover, Christina Koch, and Jeremy Hansen—four highly trained astronauts representing American and Canadian investment in human capital, international partnerships, and historic milestones— the calculus shifts toward accepting higher financial cost to reduce the probability of catastrophic failure.

Third, political and economic factors shape these decisions as much as engineering considerations. The Space Launch System sustains thousands of high-skill jobs across the United States. Manufacturing facilities in Louisiana, Alabama, Mississippi, Utah, and Florida depend on the program. Congressional representatives from those states have consistently championed SLS funding, and the 2025 One Big Beautiful Bill Act injected nearly $10 billion into NASA's budget with language mandating continued development beyond the first lunar landing. Critics call this pork-barrel spending; supporters argue it maintains an industrial base that would take decades to rebuild if allowed to collapse.

NASA's strategy is not either-or but both. The agency uses the Space Launch System for crew while stimulating commercial competition for lunar landers, cargo delivery, and eventually crew transport. By acting as an anchor customer—guaranteeing purchases for services like lunar landings and spacewalk suits—NASA helps de-risk technologies that private companies can later commercialize. Over time, competition may drive costs down. SpaceX's Starship Human Landing System and Blue Origin's Blue Moon lander are both being developed under fixed-price contracts, shifting financial risk to the companies. If these systems mature and prove reliable, NASA could eventually transition away from government-built rockets. But until that happens, the Space Launch System provides the assured access to deep space that the Artemis program requires.

The comparison with Apollo is instructive. Saturn V was an extraordinary achievement, but it was also unsustainable. Once the Apollo program ended, the rocket was abandoned and the infrastructure dismantled. Artemis aims to avoid that fate by building systems that can evolve. The Exploration Upper Stage will increase payload capacity. New production RS-25 engines use modern manufacturing techniques to reduce cost. Future boosters may incorporate advanced propulsion. If the program continues, the per-launch cost should gradually decline as production ramps up and efficiencies are found. If it doesn't continue, the cost per mission will remain high and the investment will look increasingly difficult to justify.

For readers who believe NASA should simply wait for commercial vehicles to mature, the counterargument centers on timing and competition. Waiting another five to ten years for Starship or New Glenn to demonstrate human-rating cedes ground to other nations. If the United States delays too long, it may arrive second—and in the geopolitical calculus of space exploration, firsts matter. Whether that justifies spending $4.1 billion per launch is ultimately a question of priorities: how much is leadership in deep-space exploration worth, and what are the consequences of stepping back? For Artemis II specifically, the justification is validation—proving that every dollar invested has purchased systems that work, that the crew can fly safely to the Moon and back, and that the infrastructure exists to support sustained exploration.

The Integrated Machine

Think of the Artemis vehicle stack as a long-haul freight truck engineered for a single, high-stakes journey carrying cargo worth $4.1 billion to deliver safely. The Space Launch System core stage and boosters are the massive diesel engine and fuel tanks, providing brute force to get a fully loaded vehicle onto the highway. The upper stage is the transmission, accelerating the truck to cruising speed. The Orion crew module is the driver's cab, housing Wiseman, Glover, Koch, and Hansen with seats, controls, and life-support equipment. The European Service Module is the sleeper unit and auxiliary power, containing the generator, fuel, water, and everything needed to sustain the drivers on a cross-country haul.

Each component must perform flawlessly for the trip to succeed. Unlike a truck, there are no rest stops, no roadside assistance, no chance to pull over and fix a problem. Everything must be tested, redundant systems installed, and contingencies planned. The vehicle isn't optimized for cost efficiency—it's optimized for bringing the crew home alive. That design philosophy permeates every decision, from the choice of heritage hardware to the conservative flight profiles to the exhaustive ground testing. The result is expensive but capable, a machine built to meet the demands of deep space in an era when such achievements have become rare.

When the rocket ignites and begins its climb, Wiseman, Glover, Koch, and Hansen will ride atop this machine trusting that every component will perform as designed. The 8.8 million pounds of thrust exists to accelerate them fast enough to escape Earth's gravity. The European Service Module's 11 kilowatts of power will keep their life support running for ten days. The guidance and navigation systems will keep them on course around the Moon. The heat shield will protect them through the 2,760-degree Celsius inferno of the lofted entry profile. And if any single component fails critically, the crew has trained for contingencies—abort modes, manual overrides, emergency procedures—that give them the best possible chance of survival.

The stack represents decades of engineering expertise, billions in investment, and the commitment of thousands of people across multiple nations who believe that sending humans back to deep space matters enough to accept the cost and risk. Whether that commitment proves justified depends on what Artemis II accomplishes and what follows. But on launch day, when those four astronauts strap into their seats and the countdown reaches zero, the machine built to carry them will be the most capable deep-space transportation system humans have constructed since Apollo ended—and unlike Apollo's Saturn V, it's designed to fly not once but repeatedly, building the pipeline toward sustained exploration beyond Earth.

Artemis II Mission Map
(Graphic courtesy of NASA)

CHAPTER FOUR

The Mission

Understanding what happens during Artemis II helps contextualize why each system matters and what the crew will experience during their ten days beyond Earth orbit. This chapter walks through the mission in three phases: the intense first day as the spacecraft climbs away from Earth and the crew validates critical systems; the multi-day journey to the Moon and back; and the fiery return through the atmosphere using an entry profile that has never been flown with hardware. Think of it as a guided tour through the mission timeline, emphasizing not just what happens but why each test matters for the program's survival and what it will feel like for the four astronauts strapped inside Orion.

Every validation the crew performs justifies a portion of the mission's $4.1 billion cost. If proximity operations fail to prove that astronauts can manually fly Orion during critical maneuvers, the entire Gateway architecture collapses—future missions can't dock with the lunar station or rendezvous with landers. If life-support systems struggle to handle actual crew metabolic loads, discovering that now rather than during Artemis III's lunar landing could save the program from catastrophic failure. If the lofted entry profile doesn't perform as modeled, NASA's calculated bet on choosing known risk over unknown char loss will have failed, potentially grounding the program while engineers redesign the return trajectory. The stakes are extraordinary, and the crew knows it.

For readers planning to watch the launch and follow along with NASA's coverage, this chapter also serves as a reference. The included timeline table and jargon decoder will help translate what you're hearing from Mission Control into what's actually happening aboard the spacecraft. The mission may not unfold exactly as described—spaceflight rarely does—but these are the major beats NASA expects and the challenges the crew has trained to handle.

Phase 1: Launch

The countdown reaches zero. Four RS-25 engines ignite at the base of the core stage, building thrust while computers verify that each is operating normally. When sensors confirm full power, the solid rocket boosters light with a concussive roar that shakes the launch pad and rattles windows miles away. Eight point eight million pounds of thrust overcomes gravity and inertia. The 2.6-million-kilogram vehicle begins rising, slowly at first, then with accelerating velocity as propellant burns away and the rocket becomes lighter.

Inside Orion, Commander Reid Wiseman and pilot Victor Glover feel the vibration and acceleration building. Their displays show engine performance, trajectory, and abort modes. Mission specialists Christina Koch and Jeremy Hansen watch systems that will become critical once the rocket reaches orbit: the European Service Module's power and propulsion status, the life-support parameters, communication links with ground controllers. The ride is rougher than a shuttle launch—those solid boosters create more vibration—but smoother than Apollo crews experienced. For two minutes the boosters burn, then separate in a shower of explosive bolts and small rockets that push them clear of the core stage. Acceleration eases momentarily, then builds again as the RS-25 engines continue firing.

Eight minutes after liftoff, the main engines cut off. The crew experiences a sudden shift from acceleration to weightlessness—the sensation astronauts describe as going from being pressed into your seat to floating against your straps. The core stage separates, falling away toward the ocean far below. The Interim Cryogenic Propulsion Stage ignites, its single RL-10 engine pushing Orion into an elliptical orbit with a low point near Earth and a high point around 1,200 nautical miles up—nearly five times higher than the International Space Station.

The crew has no time to adjust. Within fifty minutes of liftoff, the upper stage fires again to raise the orbit even higher, eventually establishing a 23.5-hour elliptical path that takes Orion out to roughly 38,000 miles from Earth. At that altitude, the astronauts will see the entire disk of Earth suspended in blackness—a view no human has witnessed since Apollo 17 in 1972. But they're too busy to marvel. Wiseman and Glover begin an intricate seat swap, unbuckling and floating to opposite positions so Glover can prepare for the manual flight demonstration scheduled for later in the day.

Meanwhile, Koch and Hansen power up the spacecraft's systems. Solar arrays deploy from the European Service Module, unfurling like the petals of a mechanical flower. The arrays track the sun, generating electrical power and charging batteries. The crew module's life-support system transitions from launch mode—where it provided minimal air circulation and relied on bottled oxygen—to on-orbit mode, where it must scrub carbon dioxide, regulate humidity, and recycle every breath. Koch monitors the transition closely, watching displays that show oxygen partial pressure,

CO2 concentration, humidity levels, and temperature. Her 328 days on the space station taught her what normal looks like and what subtle variations signal developing problems. In deep space, those variations matter more—there's no quick return to Earth if systems begin degrading.

Laptops boot up. Cameras are unpacked from storage lockers. The crew checks in with Mission Control via the Deep Space Network, confirming that communications work at this unprecedented distance. Every checklist item completed, every system activated successfully, represents validation that years of development and billions in investment have produced hardware that actually works.

ARTEMIS II Piloting Demonstration Test

Astronauts will pilot Orion during a test called the proximity operations demonstration. The test will check out Orion's handling qualities using the interim cryogenic propulsion stage (ICPS) as a target and inform future mission planning.

1. Orion detaches from the ICPS and does an automated flip to face its target.

2. At about 300 feet away, the crew takes controls for initial checkouts of manual capabilities, including rotation and side to side movement.

3. The crew pilots Orion to approximately 30 feet of its target.

4. The crew stops Orion and checks out fine handling qualities at close range.

5. Orion backs away and the ICPS turns to protect its thermal properties.

6. The crew initiates a second round of manual maneuvers, this time using a target on the side of the stage.

7. Orion performs a departure burn to continue on with the rest of the mission.

Piloting Demonstration Test Graphic
(Graphic courtesy of NASA)

Proving Manual Flight

About three hours into the mission comes the most demanding task of the first day, and arguably the most critical validation of the entire mission. Orion separates from the spent upper stage, backing away slowly under automated control. Then Glover takes over, using hand controllers to fly the spacecraft manually. He pitches Orion through a 180-degree flip, rotates it, and maneuvers to within ten meters of the upper stage's nose—close enough to see details on the engine bell and plumbing. He circles the stage in both directions, evaluating how Orion responds to his commands. Wiseman monitors fuel consumption, thruster temperatures, and navigation sensors, calling out any anomalies.

This test isn't about demonstrating that Glover can fly a spacecraft—his Crew Dragon experience already proved his piloting skills. It's about validating that Orion's fly-by-wire system translates human commands into precise thruster firings in a way that feels natural and predictable to pilots. If the spacecraft responds sluggishly, or if thruster authority feels inadequate, or if the handling characteristics differ significantly from simulator predictions, NASA must refine the software before future missions attempt far more demanding maneuvers.

The stakes extend far beyond this single flight. Artemis III requires precise rendezvous and docking with SpaceX's Starship Human Landing System in lunar orbit. Gateway operations will demand repeated docking approaches by Orion and cargo vehicles. If manual control proves awkward or unreliable during benign conditions—circling a passive, predictable upper stage in high Earth orbit—attempting those maneuvers in lunar orbit with real mission objectives becomes unconscionably risky. The entire architecture depends on astronauts being able to assume manual control when automation fails or when human judgment is required for complex proximity operations.

Glover approaches the demonstration methodically. He's not trying to execute perfect maneuvers—he's trying to understand the vehicle. When he commands a rotation, how quickly does it respond? When he arrests that rotation, does it stop cleanly or does momentum carry it past the desired attitude? When he fires translation thrusters to move laterally, does the spacecraft track purely sideways or does it induce unwanted rotation? These subtle characteristics matter enormously. A spacecraft that feels "twitchy" or unpredictable creates cognitive load that distracts pilots during critical moments. A vehicle that responds intuitively allows pilots to focus on strategy rather than fighting the controls.

He maneuvers close to the upper stage—within ten meters, closer than most pilots would feel comfortable without extensive practice. He holds position, feeling how the thrusters compensate for Orion's momentum and the weak gravitational perturbations at this altitude. He backs away, approaches from a different angle, circles in the opposite direction. Wiseman calls out propellant margins—they have enough for two hours of this testing, but they must save margin for unexpected situations. Koch and Hansen monitor navigation sensors, ensuring that Orion's computer maintains accurate awareness of position and orientation relative to the target.

After nearly two hours, Glover hands control back to the autopilot and the spacecraft backs farther away from the upper stage, which will eventually tumble back toward Earth and burn up in the atmosphere. The data collected during this demonstration will be analyzed for months. Engineers will compare actual thruster firing sequences against predicted ones. They'll evaluate propellant consumption, thermal control performance during sustained firing, and how accurately the navigation sensors tracked relative motion. They'll incorporate Glover's subjective feedback—what felt right, what felt wrong, what surprised him—into simulator updates so future crews can train on more accurate models.

If this demonstration had revealed that manual control was inadequate or unreliable, the entire Artemis timeline would have shifted. Gateway docking couldn't proceed without confidence that crews can fly the spacecraft precisely. Lunar landing missions would be delayed while NASA either redesigned Orion's control system or developed alternative approaches to rendezvous. The validation Glover provides in these two hours justifies a substantial portion of the mission's

$4.1 billion cost—because discovering a problem now, when it can be fixed before operational missions, prevents far more expensive failures later.

Settling Into Deep Space

Only after the proximity operations demonstration does the crew have time to catch their breath, stow their spacesuits, and reconfigure the cabin for living. They fold away unused seats, unpack exercise equipment, and set up the small galley where they'll heat meals. Glover deliberately over-exercises during this period, breathing hard and sweating, to stress the life-support system and ensure it can handle worst-case metabolic loads. Koch monitors carbon dioxide levels, humidity, and temperature, watching how quickly the system recovers. She also practices switching between suit mode and cabin mode, verifying that if they needed to don pressure suits quickly during an emergency, the environmental controls would transition smoothly.

The flight plan includes several four-hour sleep periods interspersed with tasks. But sleep on the first night is difficult. Adrenaline still courses through the crew's veins. The cabin is noisy—pumps, fans, and thrusters create a constant background hum. And the view out the windows is mesmerizing: Earth rotating slowly below, the Moon visible in the distance, stars unblinking in the absence of atmosphere. The crew eventually drifts off in sleeping bags tethered to cabin walls, adapting to a rhythm that will carry them through the rest of the mission.

Opposite: Artemis I Flight Day 13: Orion, Earth, and Moon
(Photo courtesy of NASA)

Phase 2: Moon and Back

Roughly twelve hours into the mission, Orion's main engine fires for a perigee-raise burn, circularizing the high elliptical orbit and setting up the geometry for translunar injection. The burn is automated, but the crew monitors propellant flow, engine performance, and trajectory updates closely. Once complete, they have a brief window to rest and eat before the final decision point: whether to commit to the Moon. If any systems have shown anomalies during the checkout period, Mission Control could wave off and keep Orion in Earth orbit for another day or two. But if all is well, the green light comes.

The translunar injection burn is the mission's commitment point. Once Orion's main engine fires for fifteen to eighteen minutes, accelerating the spacecraft to lunar transfer velocity, there's no easy way back. The crew is committed to a free-return trajectory—a figure-eight path that uses the Moon's gravity to sling them back toward Earth even if the engine fails. As the burn progresses, the astronauts watch their velocity climb on the displays. When the engine cuts off, they're traveling faster than any humans since Apollo, outbound toward a destination 240,000 miles away.

This burn represents the culmination of years of preparation and billions in investment reaching its moment of truth. If the European Service Module's main engine performs flawlessly, if the propellant tanks feed reliably, if the guidance system calculates the burn precisely, the crew will be on course for their lunar flyby. Any significant underperformance or off-nominal behavior could require contingency planning—aborting the lunar flyby, extending the mission duration to fix trajectory errors, or in worst case, executing emergency return procedures. The confidence NASA has in this burn comes from exhaustive ground testing and Artemis I's successful uncrewed demonstration, but until the engine fires with crew aboard, some uncertainty remains.

The Journey Out

The days that follow have a different rhythm. Without the frenzy of launch and the urgency of proximity operations, the crew settles into a routine. They perform health monitoring: checking blood pressure, collecting biological samples for the AVATAR experiment, and testing cognitive performance using computerized assessments. They calibrate navigation sensors, photograph Earth and the Moon, and conduct educational downlinks with classrooms around the world. They exercise on the resistance harness, eat meals together, and sleep in shifts. The spacecraft coasts, making only small thruster firings to correct its trajectory.

Communications shift from near-Earth ground stations to the Deep Space Network. Conversations now include a noticeable delay—several seconds for signals to travel between Earth and Orion and back. The crew learns to pause after speaking, waiting for responses. They practice communicating during simulated problems, where the delay forces them to think through solutions independently rather than relying on instant guidance from Houston. When Orion passes behind the Moon, radio contact is lost entirely for periods approaching an hour.

These communication blackouts test a capability essential for Mars missions, where delays stretch to twenty minutes round-trip and crews must operate autonomously for months. During blackouts, Orion follows pre-programmed sequences. The crew monitors systems and prepares to take manual control if needed, but mostly they wait and trust that the software will perform as designed. Wiseman's experience as chief astronaut, where he prepared crews for scenarios where they couldn't rely on immediate ground support, becomes invaluable. He maintains calm situational awareness, verifies that his crew knows what to do if anomalies arise, and models the confidence required to function effectively when Earth's voice goes silent.

Koch uses these quiet periods to focus on the AVATAR experiment—the organ-on-a-chip technology carrying bone marrow cells from each crew member. The experiment operates autonomously, but she photographs the hardware, verifies that temperature and perfusion remain nominal, and updates the experiment log. The data these chips provide about how deep-space radiation affects bone marrow at the cellular level will inform radiation protection strategies for future long-duration missions. It's unglamorous science—tending to tiny chips sealed in protective housing—but it represents exactly the kind of biological research that requires human presence in deep space.

Lunar Flyby: The Moment Everything Aims Toward

Around four days into the mission, Orion reaches its closest approach to the Moon—about 7,400 kilometers above the surface. This is not a lunar orbit; the spacecraft won't slow down enough to be captured by the Moon's gravity. Instead, it will swing around the far side on its free-return arc, using the Moon's gravitational pull to bend its trajectory back toward Earth. As Orion rounds the far side, the crew witnesses something only 24 humans before them have seen: Earth rising above the lunar horizon.

The view is stark and breathtaking. The Moon's far side is rugged, heavily cratered, untouched by the volcanic flows that smoothed parts of the near side. No human structures mar the landscape. Sunlight rakes across crater walls at low angles, creating dramatic shadows. And beyond the Moon, suspended in the blackness, Earth appears as a fragile blue marble. Hansen, representing Canada's partnership in this endeavor, takes a moment to absorb what it means to be the first Canadian to witness this view—a perspective that transforms how you understand humanity's place in the universe.

The crew spends precious minutes photographing the scene, knowing that the images will inspire millions back home, but also taking time to simply look—to absorb what it means to be this far from everything humanity has ever known. Glover reflects on what his presence here means for young people of color watching from Earth. Koch considers the generations of women who were excluded from this frontier and the symbolic weight of being the first to cross this threshold. Wiseman, as commander, feels the responsibility of bringing his crew home safely while also recognizing that this moment—this view—justifies a substantial portion of why the mission matters.

Then the Moon falls behind, and the long journey home begins.

The Coast Home and Final Preparations

The spacecraft continues coasting, now falling back toward Earth under the influence of gravity. The crew performs experiments, monitors systems, and prepares for the final challenge: reentry. They review contingency procedures for cabin depressurization, manual reentry, and off-nominal landings. They stow loose equipment, secure experiments, and reconfigure the cabin for the high-G environment of atmospheric entry. The tone shifts from the relaxed routine of deep-space cruise to the focused intensity of mission-critical operations.

Koch leads the crew through final life-support checks. They verify that carbon dioxide scrubbers have maintained performance throughout the mission, that water recycling has functioned as expected, and that oxygen generation systems have enough capacity for the final days. They document any anomalies—unusual readings, equipment that performed differently than predicted, procedures that proved awkward in practice.

This operational feedback becomes part of NASA's database, informing improvements for future missions.

Wiseman and Glover review the entry, descent, and landing procedures. They rehearse the steps: jettisoning the European Service Module, orienting the heat shield forward, monitoring trajectory during entry, responding to off-nominal situations where the spacecraft might come in too steep or too shallow. They practice the calls they'll make to each other—crisp, professional communication that maintains situational awareness even under stress. They know that once plasma forms around the capsule, they'll lose communications with Earth for several minutes. During that blackout, they must function as an independent crew, making decisions without waiting for Houston's guidance.

Phase 3: Coming Home

Reentry begins about ten days after launch. Orion's trajectory has been fine-tuned through small course-correction burns to ensure the spacecraft hits a narrow entry corridor in Earth's atmosphere. Too steep and the deceleration forces could exceed safe limits; too shallow and the capsule might skip off the atmosphere entirely, careening back into space with no way to return.

What makes this reentry uniquely challenging is that no hardware has ever flown the lofted entry profile Artemis II will use. The crew's confidence comes from analysis, simulation, and decades of expertise—but not from prior flight experience with this specific trajectory. When NASA discovered the heat shield gas-trapping issue on Artemis I, the decision to switch to a lofted entry traded the unknown risk of asymmetric char loss for the known, quantifiable risk of higher peak temperatures and G-loads. Now, as the crew prepares to execute that entry, the calculated nature of that engineering choice becomes viscerally real.

The European Service Module, having done its job, is jettisoned a few hours before entry. It will burn up in the atmosphere, its aluminum structure melting and vaporizing as friction converts the spacecraft's kinetic energy into heat. The crew module continues alone, its heat shield oriented forward.

At an altitude of about 122 kilometers—the arbitrary boundary where space begins—Orion encounters the first wisps of atmosphere. Friction begins to slow the spacecraft, and the heat shield starts to glow. The lofted entry follows a single continuous descent path, steeper and more aggressive than the skip entry Artemis I flew. The spacecraft enters at a sharper angle, experiences higher peak heating, but spends less time at intermediate temperatures where gas trapping could occur.

The heat shield does its work brilliantly—or so the models predict. As the spacecraft plows deeper into the atmosphere, temperatures at the shield's outer surface climb to around 2,760 degrees Celsius. The Avcoat material chars and ablates, carrying heat away from the underlying structure. Inside the crew module, the temperature remains comfortable—the crew feels warmth but not scorching heat. What they do feel is deceleration. The spacecraft sheds velocity rapidly, subjecting the astronauts to several times the force of gravity. They're pressed into their seats, breathing becomes harder, and vision tunnels slightly as blood is pulled toward their feet.

For several minutes, reentry plasma surrounds the capsule and blocks all communications. This blackout period is expected, but it's unnerving nonetheless. The crew has no contact with Mission Control, no confirmation that ground stations are tracking them, no reassurance that recovery forces are ready. They rely on their training and on the spacecraft's ability to navigate autonomously. Wiseman maintains calm, monitoring displays that show altitude, velocity, and G-loads. If readings deviate significantly from predicted values, he's prepared to take manual control—though at this point in the entry, options are limited.

The plasma clears and communications resume. There's palpable relief in both the capsule and Mission Control. The lofted entry has worked—the heat shield performed as modeled, temperatures remained within acceptable limits, and the crew is safely through the worst thermal environment. NASA's calculated bet on choosing known risk over unknown char loss has proven correct. This validation is worth billions in itself—every future Artemis crew will return using this profile, and the confidence now comes from actual flight data rather than simulation alone.

As Orion descends through 25,000 feet, the first parachutes deploy. Two small drogue chutes stabilize the spacecraft and slow it further. At 8,000 feet, the drogues are cut away and three main parachutes deploy, each 116 feet in diameter. The main chutes slow Orion to about 20 miles per hour—fast enough that splash-down will feel like hitting a wall, but survivable. The crew braces. The capsule slams into the Pacific Ocean with a jolt that rattles equipment and tests the strength of harnesses. Then it's over. The spacecraft bobs in the waves, temporarily inverted until airbags inflate to right it.

Recovery and Validation

Recovery forces have been staged nearby—U.S. Navy ships with teams trained specifically for Orion retrieval. Divers deploy from helicopters, attaching a collar around the capsule to stabilize it and prevent it from sinking if the hull is compromised. Small boats approach, and the crew opens the side hatch. After ten days in microgravity, standing and walking will be difficult—muscles have weakened, balance is off, and blood pressure regulation has changed. But they're home.

Orion - Dec. 11, 2022 off the coast of Baja California.
(Photo courtesy of NASA/Kim Shiflett)

For those watching from Earth, splashdown marks the end of the mission. For NASA, it's the beginning of another phase: examining every system, reviewing telemetry, interviewing the crew, and applying lessons learned to Artemis III. The questions will be exhaustive:

- Was the lofted entry profile effective? Did temperatures and G-loads match predictions? Did the heat shield perform uniformly, without the asymmetric char loss that plagued the skip entry?

- How did life-support systems handle ten days of continuous operation? Were there any degradations in performance? Did carbon dioxide scrubbing, water recycling, and thermal control maintain nominal conditions throughout?

- What did the crew learn about flying Orion manually? Did Glover's proximity operations demonstration reveal any handling characteristics that require software refinement? Can NASA confidently certify that future crews can dock with Gateway and lunar landers?

- How well did the crew function autonomously during communication blackouts? Could they make informed decisions without real-time ground support? Does the operational model scale to Mars distances where delays are far longer?

Every answer feeds into preparations for Artemis III. If the lofted entry worked flawlessly, the profile becomes standard for all future missions. If life support performed nominally, NASA gains confidence that crews can sustain themselves for the weeks-long surface missions planned for later flights. If Glover's manual flight data validates that Orion handles predictably, the Gateway rendezvous timeline can proceed. If any systems underperformed or behaved unexpectedly, engineers will analyze why and make corrections.

The mission's success ultimately determines whether the $4.1 billion investment was justified and whether the program can proceed to lunar landing. Technical success alone isn't sufficient—the political and fiscal environment demands that Artemis demonstrate value, capability, and progress toward stated objectives. But technical success is necessary, and Artemis II's primary purpose is validating that the systems work with crew aboard, that calculated engineering choices like the lofted entry prove sound, and that the pathway to sustained lunar exploration remains viable.

When Wiseman, Glover, Koch, and Hansen climb out of the capsule and onto the recovery ship, they'll have become the first humans to venture into deep space in more than fifty years. They'll have tested systems that future crews will depend on. They'll have proved that diverse, international teams can function seamlessly far from Earth. And they'll have validated NASA's most consequential engineering decision in the program— choosing to fly a new entry profile rather than delay the mission by eighteen months. Whether that validation proves sufficient to sustain political support and secure funding for Artemis III and beyond won't be known immediately. But the foundation will have been laid.

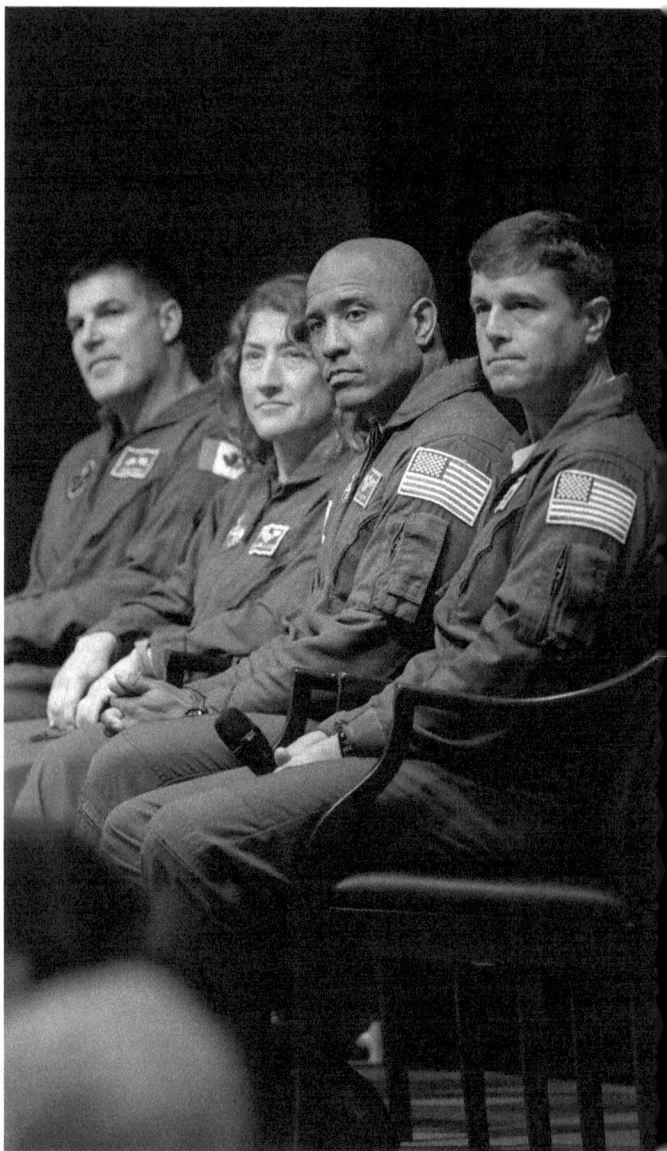

Artemis II astronauts Victor Glover, Reid Wiseman, Christina Koch, and Jeremy Hansen.
(Photo courtesy of NASA/Charles Beason)

CHAPTER FIVE

Human Validation

You've met the crew: Commander Reid Wiseman, pilot Victor Glover, mission specialists Christina Koch and Jeremy Hansen. You understand their backgrounds—military aviation, long-duration spaceflight, engineering expertise, international partnership. But biographical credentials don't prepare anyone for manually flying a spacecraft around a spent rocket stage at 38,000 miles altitude, or managing life support in nine cubic meters for ten days, or executing a reentry profile that's never been flown with hardware.

Preparation for Artemis II consumed years of their lives—thousands of hours in simulators, underwater tanks, geology field trips, and survival training. The goal wasn't to script every moment of the mission. That's impossible. Instead, training built mental models of how systems work, internalized procedures so they become automatic under stress, and developed the team cohesion required to solve problems collaboratively when things go wrong. By the time Artemis II reaches the launch pad, the crew will have flown the mission so many times in simulators that the real thing may feel familiar—until the engines ignite and the vehicle actually starts shaking, and the stakes suddenly become real.

The Human Factor

An obvious question arises: if robots and uncrewed spacecraft can orbit the Moon and return data, why risk human lives? The answer lies in what humans bring that automation cannot replicate.

Certain aspects of spacecraft performance can only be validated with people aboard. Life-support systems must respond to the unpredictable metabolic patterns of living, breathing, sweating humans—something no simulation fully captures. When Koch exercises vigorously and then rests, when Glover's respiration rate spikes during the manual flight demonstration, when Hansen sleeps while Wiseman remains awake monitoring systems, the Environmental Control and Life Support System experiences load variations that ground tests with mannequins cannot replicate. The system must adapt in real time, and engineers need data on how it performs under actual crew conditions before committing to the weeks-long surface missions planned for Artemis III and beyond.

Manual control validation requires human pilots. Autopilot algorithms must gracefully hand control to astronauts when sensors disagree or unexpected situations arise. User interfaces must remain intuitive when crews are fatigued, stressed, or working through problems that ground controllers cannot immediately advise on. During the proximity operations demonstration, when Glover evaluates whether Orion's fly-by-wire system feels natural or awkward, he's providing feedback that only a pilot with hands on the controls can give. Does the spacecraft respond crisply or sluggishly? Do thruster firings feel proportional to stick inputs? Does the navigation display provide the information he needs at the moment he needs it? These subjective assessments inform software refinements that make future docking operations with Gateway and lunar landers safer and more reliable.

Humans make decisions that automation struggles with. When communications lag several seconds because of distance, or when the spacecraft passes behind the Moon and loses contact entirely for an hour, the crew must troubleshoot problems independently. If a sensor disagrees with the flight computer's trajectory estimate, Wiseman can cross-check using visual references—sighting on stars, observing Earth's horizon—and make judgment calls about which data source to trust. If equipment fails in unexpected ways, astronauts can improvise solutions using whatever materials are available, as Apollo 13 demonstrated when the crew jury-rigged carbon dioxide scrubbers to survive the trip home. Automation excels at executing planned procedures; humans excel at adapting when plans prove inadequate.

The intangible dimension matters too. A spacecraft crewed by humans is fundamentally different from one that's empty. The mission becomes a shared experience that millions follow, identifying with the astronauts and feeling invested in their safe return. Students see Glover manually flying Orion and begin imagining careers in aerospace engineering. Young women watch Koch manage life support in deep space and recognize that technical expertise has no gender barriers. Canadian students follow Hansen's journey and take pride in their nation's contribution to exploration. The public engagement that crewed spaceflight generates translates into sustained political support for exploration budgets. Whether that justifies the additional cost and risk is debatable, but NASA's history shows that human missions capture attention in ways that robotic probes—no matter how scientifically valuable—typically do not.

For Artemis II specifically, humans validate that the $4.1 billion transportation system actually works for its intended purpose: safely carrying crews to deep space and back. An uncrewed test proved the hardware functions. A crewed test proves that humans can live inside it, operate it manually when required, and return safely through an entry profile born from engineering necessity rather than original design. Without this validation, attempting Artemis III's lunar landing would be unconscionably risky.

Canadian Space Agency astronaut Jenni Gibbons practices simulated lunar tasks at NASA's Neutral Buoyancy Laboratory in Houston.
(Photo courtesy of NASA/Daniel O'Neal)

Training for the Unknown

The crew's preparation began with technical immersion—learning Orion's systems inside and out. They studied propulsion, guidance, communications, life support, and power systems—not just how to operate them but how they're designed, what failure modes exist, and what indicators suggest trouble before full failure occurs. They spent weeks at contractor facilities watching hardware being built and tested, asking engineers questions that often revealed assumptions no one had challenged. This deep understanding means that when a caution light illuminates during flight, the crew doesn't just follow a checklist—they understand what's happening behind the panel and can adapt if the checklist doesn't fit the situation.

Simulator training formed the core of preparation. The crew has flown Artemis II dozens of times in high-fidelity simulators at Johnson Space Center, practicing nominal operations and responding to failures. Instructors inject malfunctions at the worst possible moments: a thruster fails during proximity operations, forcing Glover to compensate with remaining thrusters while avoiding collision with the upper stage. The life-support system develops a carbon dioxide scrubbing problem during the outbound coast, requiring Koch to troubleshoot under time pressure while her crewmates continue mission tasks. The main engine refuses to ignite for translunar injection, forcing Wiseman to decide whether to abort the lunar flyby or try a backup system. Each scenario builds problem-solving skills and reveals how the crew communicates under stress.

Glover spent hundreds of additional hours in the proximity operations simulator, learning Orion's handling characteristics. The spacecraft responds differently than Crew Dragon, which he flew to the International Space Station. Orion is heavier, the thruster authority differs, the fly-by-wire control laws translate stick inputs differently. He practiced approaches from multiple angles, circling in both directions, holding position against perturbations, and backing away smoothly. He learned what "normal" feels like so he can recognize when something feels wrong during the actual flight. His test pilot background—evaluating aircraft that haven't been flown before and articulating what works and what needs refinement—becomes critical during the demonstration that will validate manual control for all future Artemis missions.

Koch trained extensively on life-support monitoring. She learned not just the nominal displays but what subtle variations in oxygen partial pressure, carbon dioxide concentration, humidity, and temperature signal developing problems. Her 328 days aboard the International Space Station taught her that systems often show early warning signs—readings that drift gradually outside normal ranges, pumps that sound slightly different, filters that load faster than expected. Catching these early prevents small issues from cascading into emergencies. During Artemis II, she'll deliberately stress the system by varying crew activity levels, documenting how quickly it recovers, and identifying any edge cases where performance degrades. The data she collects becomes part of NASA's database on how environmental control behaves in deep space, informing Gateway design and Mars mission planning.

Hansen focused on communications systems and international coordination. He learned how to manage transitions between near-Earth ground stations and the Deep Space Network, how to communicate effectively during multi-second delays, and what procedures apply when Orion passes behind the Moon and loses contact entirely. As liaison with Canadian partners tracking the flight, he practiced explaining mission status, coordinating data sharing, and managing expectations when delays or anomalies occur. His role exemplifies the international character of Artemis—not just one nation exploring but a coalition where each partner contributes and benefits.

Integrated simulations involved not just the crew but the entire mission team—flight controllers, capsule communicators, recovery forces, international partners. These multi-day exercises replicated every phase of the mission, from pre-launch checks through splashdown. Everyone practiced their roles, learned each other's communication styles, and discovered where coordination breaks down. When a simulated problem required input from engineers in Europe who built the service module, the team worked through time zones and language barriers they'll face during the real mission. When weather forced a simulated splashdown in a different ocean zone, recovery forces practiced redirecting ships and helicopters on short notice.

The crew also trained underwater at NASA's Neutral Buoyancy Laboratory. Although Artemis II includes no spacewalks, underwater sessions built team coordination and situational awareness. They practiced maneuvering in confined spaces, managing tethers and tools while their movements were slowed by water resistance. The experience translated to the microgravity environment inside Orion, where objects float unless secured and where coordinating four people in a small capsule requires constant awareness of everyone's position.

Physical conditioning remained essential throughout preparation. The crew maintained rigorous fitness routines to ensure their bodies could withstand launch acceleration, adapt to microgravity, and handle the intense G-loads of the lofted entry profile. They practiced using the exercise harness they'll fly on the mission, learning to generate resistance that maintains muscle tone without the bulky equipment available on the space station. They studied cardiovascular changes that occur in microgravity—how blood pools in the upper body, how the heart doesn't have to work as hard to pump against gravity—and they understand the countermeasures required to minimize deconditioning during the relatively short mission.

Psychological preparation mattered as much as physical readiness. The crew worked with NASA's Behavioral Health and Performance team to build resilience for isolation and confinement. They learned self-assessment tools to monitor their own stress and mood. They practiced communicating concerns without defensiveness and resolving conflicts constructively when confined in a space the size of a minivan with no privacy and no escape. They discussed scenarios where disagreements arise—how to voice dissent without undermining leadership, how to receive criticism without becoming defensive, how to maintain team cohesion when everyone is tired and irritable. They prepared care packages from family that will be opened at predetermined points during the mission, timed to provide emotional boosts when the novelty of space-flight has worn off and the days feel long.

Perhaps most importantly, the crew built trust. They've spent so much time together—in simulators, in training facilities, on geology field trips, during survival exercises—that they've learned each other's strengths, weaknesses, communication styles, and stress responses. They know when someone's tone of voice means they're uncertain versus when it means they're confident. They've argued in simulations, sometimes sharply, and learned to resolve disagreements constructively. They've made mistakes and supported each other through the aftermath. When the mission begins and the spacecraft leaves Earth's protective embrace, that trust will be the foundation everything else builds on.

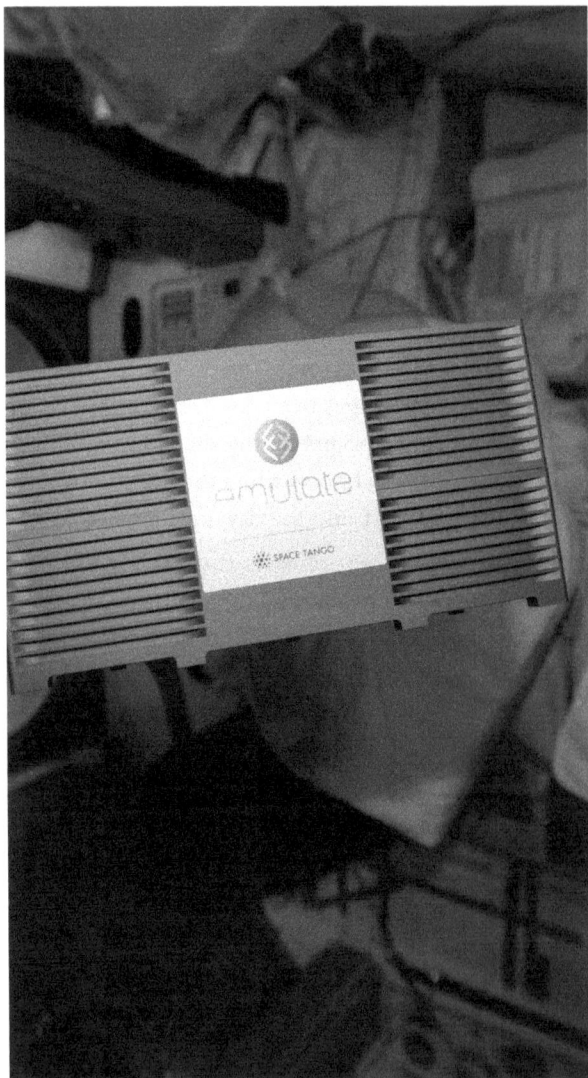

A view of the Gut on Chip CubeLab aboard the International
Space Station (ISS).
(Photo courtesy of NASA/Drew Morgan)

The Science of Deep Space

Artemis II carries experiments designed to probe how the deep-space environment affects living tissue. The most innovative is AVATAR, an organ-on-a-chip study developed by NASA and Emulate Biosciences. Before launch, researchers drew blood from each crew member and isolated bone marrow progenitor cells—the stem cells that produce blood cells and maintain the immune system. These cells were grown into miniature bone-marrow-like tissues within microfluidic chips about the size of a thumb drive, perfused with nutrients and maintained at body temperature.

One set of chips remains on Earth as a control, housed in an incubator that mimics the conditions aboard Orion. Another set flies on the mission, exposed to deep-space radiation and microgravity. The experiment essentially creates a living replica of each astronaut's bone marrow and sends it to the Moon alongside them. After the mission, researchers will compare the flight and control samples, looking for differences in gene expression, cell viability, DNA damage, and immune function. The technology reveals how deep-space radiation affects the body's ability to fight infection and heal wounds—critical knowledge for missions lasting months or years.

Koch will monitor the AVATAR hardware throughout the mission, photographing the chips, verifying that temperature and perfusion remain nominal, and documenting any anomalies in the experiment log. Her background designing scientific instruments for space missions gives her appreciation for what the technology represents: a way to study human cells directly in the environment they'll experience on Mars, without extrapolating from mouse studies or relying on radiation models that may not capture biological reality. The chips are delicate—sensitive to temperature fluctuations, vibration, and any interruption in perfusion—so her monitoring becomes part of ensuring the experiment survives to provide usable data.

The experiment also reduces the need for animal testing. Traditional radiation studies expose mice or other laboratory animals to simulated space environments, then sacrifice them for tissue analysis. Organ-on-a-chip technology allows researchers to study human cells directly, with living tissue that responds to radiation more like an actual human body than mouse tissue does. If the approach proves successful, it could accelerate medical research on Earth while providing better data for protecting astronauts on Mars missions.

Beyond AVATAR, the crew carries personal dosimeters that measure radiation exposure throughout the mission. These small devices track cosmic rays, solar particles, and secondary radiation created when high-energy particles strike the spacecraft's hull. The data will validate shielding models and inform decisions about how much radiation exposure is acceptable for different mission profiles. If solar activity increases and radiation levels spike, how much margin does the crew have before exposure limits are exceeded? How effective are the spacecraft's materials at blocking different types of radiation? The answers shape mission planning for Gateway and Mars.

The crew will also monitor their own health throughout the flight. They'll take biological samples—blood, saliva, urine—at predetermined intervals and freeze them for post-flight analysis. They'll track their cognitive performance using computerized tests that measure reaction time, memory, and decision-making. They'll document sleep quality, mood, and physical symptoms. This biomedical data becomes part of NASA's growing database on how humans adapt to space, feeding into the development of countermeasures for longer missions.

Other experiments examine how equipment behaves in deep space. The crew will test new communications protocols, evaluate user interface designs, and assess whether certain materials degrade when exposed to the thermal cycling between sunlight and shadow. They'll photograph Earth and the Moon using high-resolution cameras, providing imagery that scientists will study for decades. Every aspect of the mission—from the mundane to the groundbreaking—contributes data that makes future flights safer and more capable.

Artemis II crew training takes place at multiple NASA centers and partner facilities. At Johnson Space Center in Houston, the crew uses the Orion cockpit simulator—a full-scale replica with functional displays, controls, and motion platforms that simulate acceleration and vibration. The Neutral Buoyancy Laboratory, a 6.2-million-gallon pool, allows underwater training that approximates the microgravity experience. Kennedy Space Center in Florida hosts rehearsals with the actual flight hardware, including the Countdown Demonstration Test where the crew boards Orion and practices launch procedures.

The crew has also trained at contractor facilities: Boeing's SLS factory in Louisiana, Lockheed Martin's Orion assembly plant in Colorado, and Airbus's European Service Module facility in Germany. These visits allow astronauts to understand how hardware is built, ask engineers detailed questions, and observe testing. International partners provide training on their contributed systems—Canada's robotics, Europe's life-support components, Japan's experiments—ensuring the crew can operate or troubleshoot equipment from multiple sources.

Mission simulations integrate the entire flight operations team. These multi-day exercises replicate launch, coast phases, proximity operations, lunar flyby, and reentry. Instructors inject realistic failures at critical moments to test crew and ground team responses. After each simulation, the team conducts a debrief to identify what worked, what didn't, and what procedures need refinement. By launch day, the crew and mission control will have flown Artemis II successfully dozens of times and unsuccessfully just as often, learning from every scenario.

Ready to Fly

By the time Artemis II launches, the crew will have flown the mission so many times in simulators that certain moments will feel routine. They'll have rehearsed donning their bright orange pressure suits, riding the elevator to the top of the launch tower, and strapping into Orion's seats. They'll have practiced emergency egress using the zip-line baskets that can whisk them away from the pad if the rocket develops a problem. They've spent weeks at sea with Navy recovery forces, learning what splashdown feels like and how to exit the capsule while it bobs in ocean swells.

But no amount of training fully prepares you for the reality. The G-forces during launch will feel more intense than simulators can replicate. The view of Earth from 38,000 miles will be more stunning than photographs suggest. The silence during communication blackouts behind the Moon will be more profound than any practice session. The tension of reentry—knowing that the lofted entry profile must work or the calculated bet on known risk versus unknown char loss will have failed—will focus the mind in ways that simulations, where you can always hit reset, cannot match.

What training provides is confidence rooted in competence. When unexpected situations arise—and they will—the crew won't panic because they've solved similar problems before, sometimes repeatedly. When systems behave oddly, they'll trust their understanding of how things work and make reasoned decisions. When communication with Earth is delayed or lost, they'll operate autonomously because they've practiced that scenario until it feels routine. And when the mission is over and they splash down in the Pacific, they'll climb out of the capsule knowing they've validated not just hardware but the human capacity to venture into deep space and return safely.

The crew also understands the symbolic dimension of their flight. They're not just testing a spacecraft—they're representing a vision of exploration that's inclusive, international, and collaborative. When Wiseman commands the mission, he demonstrates that American leadership in space remains viable. When Glover pilots Orion manually, he shows young people of color that they belong in these roles. When Koch conducts experiments in deep space, she makes real the promise that barriers to exploration are coming down. And when Hansen participates as the first Canadian beyond low orbit, he affirms that the future of space belongs to partnerships, not unilateral endeavors.

They must balance those symbolic responsibilities, focusing on the technical work while remaining aware of the broader context. In interviews before the mission, they've emphasized that they're part of a team—not just the four of them, but the thousands of engineers, technicians, controllers, and support staff who make the mission possible. They've expressed gratitude for the opportunity while acknowledging the risks. And they've made clear that Artemis II is a beginning, not an ending. If they succeed, the pathway to the Moon opens. If they encounter problems, the program adapts and tries again. Either way, their flight will teach NASA and its partners what comes next, validating an architecture that must work if sustained exploration beyond Earth is to become reality.

Artemis I Space Launch System (SLS) and Orion spacecraft on Launch Pad 39B at NASA's Kennedy Space Center in Florida on Sept. 15, 202

(Photo courtesy of NASA/Jason Parrish)

CHAPTER SIX

Launch Week Playbook

This chapter is designed to be your companion during launch week and launch day itself. Unlike the previous chapters that explain hardware and mission objectives, this one focuses on what you'll actually see and hear if you're watching NASA's coverage—either from the comfort of your home or standing on a beach in Florida feeling the sound waves wash over you. It translates the technical language you'll hear from Mission Control, explains why certain events matter, and helps you understand when delays or holds are routine versus when something significant is happening.

Keep this chapter bookmarked. When the countdown reaches T-minus 10 minutes and the commentary becomes rapid-fire technical callouts, you'll want this reference to decode what's going on. When a hold is called and announcers start discussing weather or range concerns, you'll understand whether launch will happen today or whether you should settle in for a longer wait. The goal is to transform you from a passive observer into someone who can follow the technical drama unfolding and explain it to others watching alongside you.

Launch day for a crewed mission to the Moon will be unlike anything most viewers have experienced. The stakes are higher than routine cargo launches to the space station. Four astronauts are betting their lives on a $4.1 billion system designed to validate an entry profile that's never been flown with hardware, while the program's political survival depends on flawless performance. The vehicle is larger and more complex. The risks are more visible. And the emotional weight—knowing Reid Wiseman, Victor Glover, Christina Koch, and Jeremy Hansen are strapped into that capsule—adds a dimension that uncrewed flights lack.

Pre-Launch: The Days Before

Launch week actually begins weeks earlier with an event that's become a ritual for NASA's biggest missions: the rollout. The Space Launch System and Orion stack, fully assembled inside the Vehicle Assembly Building at Kennedy Space Center, begins a slow journey to Launch Complex 39B aboard a crawler-transporter—a gargantuan tracked vehicle that looks like something from a science fiction film. The journey covers four miles and takes eight to twelve hours, with the rocket moving at less than one mile per hour. It's not dramatic, but it is mesmerizing. Crowds gather along the route to watch this 322-foot-tall machine creep toward the ocean, and the sight of it silhouetted against Florida's sky reminds you just how audacious this endeavor is.

Once at the pad, the rocket is secured to the mobile launcher—a steel structure that weighs more than the rocket itself and contains the fuel lines, electrical connections, and flame ducts required to launch. Over the next several days, teams conduct checkouts. They verify that ground systems can communicate with the rocket. They test the connections between the launch tower and Orion. They power up avionics and run diagnostic routines. Much of this happens out of public view, but NASA typically provides periodic updates and photos showing the vehicle standing against the Atlantic horizon.

About a week before launch comes the Wet Dress Rehearsal, often abbreviated as WDR. This is a full

NASA's Space Launch System (SLS) rocket with the Orion spacecraft aboard is seen at Launch Complex 39B, Friday, March 18, 2022.
(Photo courtesy of NASA/Joel Kowsky)

countdown simulation where teams load the rocket with more than 700,000 gallons of cryogenic propellant—liquid hydrogen and liquid oxygen chilled to temperatures that would freeze flesh on contact. The WDR validates that ground systems can fuel the rocket safely, that propellant behaves as expected in the tank, and that the countdown timeline works as planned. Engineers watch for leaks, ice formation on fuel lines, and sensor anomalies. If problems appear, they're addressed before the crew arrives. The WDR typically proceeds to within seconds of engine ignition, then the propellant is drained and the rocket safed. It's a high-fidelity rehearsal that catches issues early, but it's also expensive and time-consuming—each WDR consumes propellant and hardware life.

After the WDR comes a quieter period where teams review data and make any necessary adjustments. The crew arrives at Kennedy about a week before launch, entering a modified quarantine protocol designed to minimize their exposure to illness. Unlike the strict isolation Apollo crews endured—two weeks with almost no outside contact—modern quarantine is more flexible. The astronauts can interact with family and support staff who've also been health-screened, but large gatherings and public events are off-limits. The goal is to prevent a common cold or flu from scrubbing the launch.

Two to three days before launch, the crew participates in a final integrated simulation called the Countdown Demonstration Test, or CDDT. They don their bright orange Advanced Crew Escape Suits—pressure garments that provide emergency oxygen and protect against cabin depressurization—and ride the elevator to the White Room at the top of the launch tower. They enter Orion through the side hatch, strap into their seats, and practice the pre-launch checklist while ground teams work through their procedures. The test verifies that communication systems work, that the crew can reach all controls while wearing bulky suits and gloves, and that emergency egress procedures function if needed. The crew doesn't stay long—a few hours at most—but the exercise builds familiarity and confidence.

The night before launch, the crew sleeps in the Neil Armstrong Operations and Checkout Building at Kennedy, the same facility where Apollo astronauts spent their final night before flying to the Moon. Tradition holds that they're served a special breakfast—steak and eggs for many crews, though menus vary—and then begin suiting up in earnest. Family members say their goodbyes, understanding that the next time they see the astronauts, those astronauts will have journeyed farther from Earth than any humans in half a century.

Understanding the Constraints

Launch windows for lunar missions are far more restrictive than for routine flights to the International Space Station. A station launch has a window measured in minutes—the rocket must lift off within a narrow time frame to match the station's orbital plane. But if the launch scrubs, another opportunity comes

24 hours later when Earth's rotation brings the launch site back into alignment. Lunar missions lack that daily cadence. The timing depends on where the Moon is in its orbit, where Earth is in its rotation, and what trajectory the mission requires. For Artemis II, these constraints combine to create launch windows that open for roughly 30 to 90 minutes on specific days, with gaps of days or weeks between opportunities.

The first constraint is the Flight Termination System battery. This safety device allows range controllers to destroy the rocket if it veers dangerously off course, protecting populated areas from a wayward missile. The FTS batteries have a certified lifespan once installed—typically around 25 to 30 days. After that, they must be removed, replaced, and recertified, which takes weeks. This creates a hard limit: if the rocket doesn't launch within the battery's certification window, the mission must stand down for extended maintenance.

The second constraint is orbital mechanics. Artemis II targets a free-return trajectory that brings the crew around the Moon and back to Earth, with splashdown occurring in daylight hours within reach of recovery forces. The trajectory depends on the Moon's position relative to Earth, which changes continuously. Only on certain days does the geometry align such that a launch from Kennedy can place Orion on the correct path. When alignment does occur, the launch window is shortened by the need to avoid certain lighting conditions at the landing site and to ensure the crew doesn't spend excessive time in the Van Allen radiation belts during the initial orbits.

The lofted entry profile adds another complication. Unlike the skip entry used on Artemis I, which provided more flexibility in landing zones, the lofted entry

has a narrower footprint. The spacecraft enters the atmosphere at a steeper angle and follows a more constrained path, which means the launch azimuth—the compass direction the rocket flies—must be tightly controlled. This further reduces the launch window. Where Artemis I might have had a two-hour window on a given day, Artemis II might have only 45 minutes.

Weather is the third major constraint, and it's more demanding for crewed missions. NASA's launch commit criteria specify acceptable limits for wind speed, temperature, lightning probability, cloud cover, and precipitation. Upper-level winds are particularly critical—strong shear at high altitudes can stress the rocket's structure during ascent or push it off course, requiring the guidance system to work harder to stay on trajectory. Surface winds matter too, especially if the crew needs to use the emergency egress system; high winds make the zip-line baskets dangerous. Precipitation concerns center on lightning risk: even light rain can create static buildup on the rocket, and lightning within ten miles of the pad during fueling operations is an automatic scrub.

Range conflicts provide a fourth constraint. The Eastern Range, managed by Space Force, serves multiple users—military launches, commercial satellites, cargo missions to the station. If another mission has priority or if tracking systems need maintenance, Artemis II must wait. Coordination among range users happens months in advance, but last-minute conflicts can arise, particularly if a previous launch encountered problems that require investigation before the range is cleared for reuse.

All these constraints compound. A perfect weather day

might fall outside the orbital mechanics window. A day with ideal geometry might have unacceptable upper-level winds. The result is that even a launch period spanning several weeks might offer only a handful of actual launch opportunities, each with a narrow window. Spectators planning to watch in person should prepare for flexibility—the first attempt may scrub, and the next opportunity might be days away.

Launch Day Viewer's Guide

Assuming weather cooperates and no technical issues arise, launch day follows a carefully choreographed timeline. NASA typically begins public commentary about four hours before liftoff, though work at the pad starts much earlier. By the time you tune in, teams have already powered up the rocket, conducted systems checks, and begun loading propellant. The commentary will orient you to where things stand and what milestones remain.

T-minus 4 hours: The closeout crew—technicians in white suits who assist the astronauts—prepares the White Room at the top of the launch tower. Inside the Vehicle Assembly Building, the crew finishes suiting up and conducts final communication checks. Family members gather for brief farewells. These moments are often shown on NASA TV, and the astronauts typically appear calm and focused, smiling for photos but clearly in work mode.

T-minus 3 hours: The crew departs the Operations and Checkout Building in a convoy of vehicles, traditionally waving to cameras as they board. The drive to Launch Complex 39B takes about 20 minutes. Upon arrival, they ride the elevator to the 300-foot level where Orion sits. One by one, they enter the capsule through the side hatch, assisted by the closeout crew who help them into their seats and verify harness connections. Commander Wiseman enters first and takes the left seat. Pilot Glover follows into the right seat.

Mission specialists Koch and Hansen take the center seats. Once all four are strapped in, technicians connect communication umbilicals, verify suit pressure, and close the hatch.

T-minus 2 hours 30 minutes: With the crew aboard, teams conduct leak checks on the hatch seal and verify that cabin pressure holds steady. The Launch Control Center polls all stations—propulsion, guidance, range, weather, recovery—for status. Each responds with "Go" or "No-go" for continuing the countdown. If all stations are "Go," the countdown proceeds. If not, holds are called to resolve issues.

T-minus 2 hours: Propellant loading intensifies. Liquid hydrogen and liquid oxygen flow into the core stage and upper stage at carefully controlled rates. The commentary will mention milestones like "core stage LOX loading" or "upper stage LH2 loading." You may see vapor clouds billowing from the rocket as cryogenic propellant boils off—this is normal. The clouds are most dramatic around the liquid hydrogen tank, where temperatures plunge to -423°F. Ice forms on the outside of the tank and falls away in sheets as the vehicle vibrates slightly from the flow of propellant.

T-minus 1 hour: The Launch Director conducts a final poll, verifying that all systems remain "Go." Weather officers provide updated forecasts for the launch window, and range officers confirm that the Eastern Range is clear. If conditions remain favorable, the countdown proceeds into terminal count—the final automated sequence that leads to ignition.

T-minus 45 minutes: The crew conducts final suit pressure checks and verifies that all communication systems work. They review abort modes, ensuring everyone understands the plan if something goes wrong during ascent. Meanwhile, ground teams prepare to retract the crew access arm—the bridge that connects the launch tower to Orion. This arm remains in place until just before terminal count begins, providing a final escape route if needed.

T-minus 20 minutes: The countdown typically enters a planned 10-minute hold at this point. This built-in pause allows teams to catch up if earlier tasks ran long and provides a buffer to address any last-minute concerns. Commentary often becomes less active during holds, with announcers filling time by reviewing mission objectives or interviewing guests. If the hold extends beyond the planned duration, that's the first signal that something may be amiss.

T-minus 10 minutes: Terminal count begins. From this point, the sequence is largely automated, controlled by computers that monitor thousands of parameters per second. The crew access arm retracts. The launch abort system arms, meaning the tower atop Orion is ready to fire if needed to pull the crew clear. Ground support equipment begins pulling back from the rocket. Cameras positioned around the pad provide stunning views of the vehicle, now glowing with ice and venting propellant vapor.

T-minus 3 minutes: The final "Go/No-go" poll occurs. This is the last moment to halt the countdown before propellant systems begin their final preparations. If any station calls "No-go," the count stops immediately and teams begin safe procedures. If all stations call "Go," the commentator will say something like "We are Go for launch," and the tension builds palpably.

T-minus 30 seconds: The automated ground launch sequencer hands control to the rocket's onboard computers. From here, only the computers or an abort command can stop the launch. Propellant valves move to their final positions. Hydraulic systems pressurize. Guidance systems complete their last alignment checks.

T-minus 10 seconds: The sound suppression system activates, flooding the flame trench beneath the rocket with thousands of gallons of water per second. This protects the pad and the rocket from acoustic damage when the engines ignite. You'll see a massive spray of water if watching from the pad area, and the roar of water hitting hot surfaces becomes audible even miles away.

T-minus 6 seconds: The four RS-25 engines ignite in a staggered sequence—engine four, then three, then two, then one—building to full thrust while computers verify that each is operating within parameters. The rocket doesn't move yet; massive hold-down bolts secure it to the mobile launcher. You'll see brilliant orange flame erupt from the base, and onlookers close to the pad will feel the air vibrate from the engines' roar.

T-minus 0: If the engines reach full thrust and all

parameters look good, explosive bolts fire to release the hold-downs and the solid rocket boosters ignite. The boosters produce a blinding white exhaust plume and a sound unlike anything else—a crackling roar that physically shakes the ground. The rocket begins to rise, slowly at first, then with accelerating speed. The commentary will call "Liftoff of Artemis II" and the moment becomes real.

For the first two minutes, the climb is steady and mesmerizing. The rocket rotates slightly, pitching toward its desired trajectory. It accelerates as propellant burns away and the vehicle becomes lighter. At about 80 seconds, you'll hear the call "Max-Q"—the point of maximum aerodynamic pressure, where the combination of speed and atmospheric density places the most stress on the vehicle's structure. The engines throttle back slightly to reduce stress, then throttle back up once Max-Q passes.

At roughly two minutes, the solid rocket boosters burn out and separate, tumbling away in opposite directions with small separation motors pushing them clear. The vehicle lurches noticeably—astronauts describe it as going from riding a freight train to a sports car—as the smoother-burning RS-25 engines continue alone. The rocket is now high enough and traveling fast enough that it looks like a bright star ascending, trailing an orange plume.

At eight minutes, the RS-25 engines cut off. The crew experiences a sudden shift from heavy acceleration to weightlessness. The core stage separates, and the upper stage ignites, pushing Orion higher. From this point, most of the vehicle is out of sight from ground observers, though tracking cameras may catch glimpses. The commentary shifts to telemetry read-outs—altitude, velocity, trajectory—as Mission Control monitors the spacecraft's climb to orbit.

About ten minutes after liftoff, the upper stage reaches orbit and cuts off. The crew is in space, though still far from their destination. Over the next several hours, they'll be performing maneuvers, raising their orbit, checking out systems, and preparing for the proximity operations demonstration. But for viewers watching from Earth, the visible part of launch is over. The rest of the mission unfolds through telemetry, radio communications, and occasional video down-links from the crew.

When Things Go Wrong

Scrubs and delays are frustratingly common in spaceflight, and Artemis II will be no exception. If you're planning to watch the launch, prepare yourself mentally for the possibility that nothing will happen on the first attempt. The odds of launching on the first opportunity are surprisingly low—weather violations, technical issues, or range conflicts can each independently halt the countdown. Understanding why delays occur and what they mean helps manage expectations and reduces disappointment.

Weather scrubs are the most common and the least worrying. If upper-level winds exceed limits, or if lightning is detected within ten miles, or if cloud cover violates criteria, the launch director will call a scrub. These decisions are made relatively late—sometimes within the final hour of the countdown—as weather officers continuously update forecasts. A weather scrub typically means the next attempt will occur when forecasts improve, which might be the following day or might be several days later depending on orbital mechanics and the launch window availability.

Technical scrubs occur when hardware doesn't behave as expected. A valve might fail to close completely. A sensor might give an anomalous reading. A software handshake between ground systems and the rocket might timeout. Often these issues are minor and can be resolved quickly, but the conservative approach to crewed flight means that any uncertainty results in a hold while teams investigate. If the problem can be fixed within the countdown timeline, the launch proceeds. If not, the attempt scrubs and teams assess whether the issue can be resolved in time for the next launch opportunity or whether longer maintenance is required.

Holds during the countdown come in two flavors: planned and unplanned. Planned holds occur at predetermined points—typically T-minus 4 hours, T-minus 20 minutes, and sometimes T-minus 10 minutes—and provide buffers in the timeline to catch up if earlier tasks ran long. These are normal and expected. Unplanned holds happen when something unexpected occurs: a boat drifts into the restricted zone downrange, requiring range officers to halt the count until it's cleared. A ground system throws an error, requiring technicians to verify it's a false alarm. A sensor reading doesn't match expectations, requiring engineers to review telemetry.

The key to understanding holds is listening to the commentary. If the hold is brief and announcers indicate teams are working through a checklist or waiting for a ship to clear the area, the launch will likely proceed once the issue resolves. If the hold extends beyond thirty minutes with no clear resolution timeline, or if teams are discussing replacing hardware or conducting additional tests, the probability of a scrub increases. Experienced launch watchers develop an ear for the tone of voice in Mission Control—relaxed and routine suggests minor issues; terse and urgent suggests something more serious.

Abort modes are what happens when something goes critically wrong during ascent. NASA has designed multiple abort scenarios depending on when a failure occurs. In the first roughly 30 seconds after liftoff, if the rocket experiences a catastrophic failure, the Launch Abort System fires. Eight solid rocket motors at the top of the Orion abort tower ignite, producing 400,000 pounds of thrust and yanking the crew capsule away from the failing rocket at accelerations exceeding 15 G's—enough to cause pain but not injury. Thrusters spin the capsule to point the heat shield forward, then the tower is jettisoned and parachutes deploy to bring the crew down on land or in the ocean. The entire sequence takes seconds and is fully automated. If you see the abort tower fire during launch, something has gone catastrophically wrong, but the crew should survive.

Later in ascent, other abort modes activate. If an engine fails but the rocket remains controllable, the vehicle can continue to orbit using remaining engines or attempt an Abort to Orbit, reaching a lower-than-planned orbit where the crew can assess their situation. If too many engines fail, an Abort Once Around returns the crew to Earth after one orbit, splashing down in the Pacific roughly 90 minutes after launch. These modes are contingencies drilled extensively in training but never yet used on a crewed U.S. mission—Apollo and Space Shuttle flew for decades without invoking them. Still, they exist as insurance, and understanding them helps contextualize the risks.

When Wiseman, Glover, Koch, and Hansen strap in, they know the abort modes intimately. They've practiced them in simulators, understand the accelerations they'll experience, and trust that the systems will work if needed. The crew will return using a lofted entry profile that's never been tested with hardware—a calculated engineering choice born from Artemis I's heat shield discovery. If that profile performs as modeled, future crews will use it confidently. If anomalies arise, NASA will adapt again, refining the approach before committing to lunar landings.

For viewers hoping to witness history, the best advice is patience and flexibility. If you're traveling to Florida to watch in person, allow extra days in your schedule. Book refundable accommodations if possible. Monitor NASA's website and social media for official updates—rumors and speculation proliferate online, but only NASA's announcements are authoritative. And remember that a scrub is not a failure. It's a demonstration that safety protocols work and that teams won't launch until conditions are right. The mission will fly when it's ready, whether that's on the first attempt or the fifth.

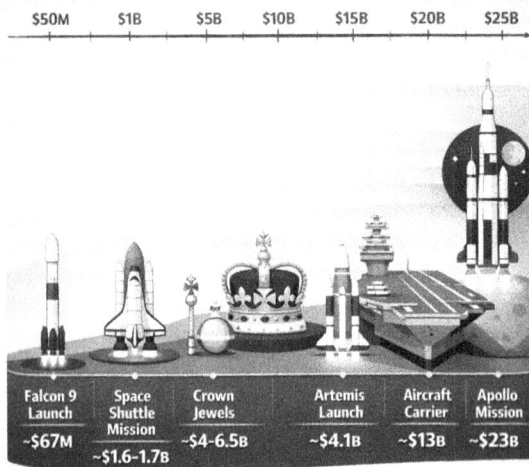

What Does a Billion Dollars Buy?

$50M	$1B	$5B	$10B	$15B	$20B	$25B

Falcon 9 Launch	Space Shuttle Mission	Crown Jewels	Artemis Launch	Aircraft Carrier	Apollo Mission
~$67M	~$1.6–1.7B	~$4–6.5B	~$4.1B	~$13B	~$23B

Values are approximate and inflation-adjusted to 2020 USD. Estimates vary by methodology.

"What does a Billion Dollars Buy" Infographic
(Infographic by MS Designer)

CHAPTER SEVEN

What Nearly Killed Artemis

You've heard the $4.1 billion figure throughout this book. It appeared when discussing why each Artemis II system validation matters—if proximity operations fail to prove manual flight capability, that investment produces nothing useful for Gateway docking. It surfaced when explaining NASA's heat shield decision—choosing to fly a new entry profile rather than delay eighteen months and miss the competitive window with China. It framed why life-support testing carries such weight—discovering problems now, before committing to lunar landings, could save the program from catastrophic failure.

Now here's the complete story behind those numbers: how the Artemis program reached this cost, why it's survived multiple attempts to cancel or scale it back, what oversight bodies say about its management, and whether the investment can be sustained through the 2030s.

The goal is clarity, not advocacy. Artemis is expensive, controversial, and fragile. It's also ambitious, inspiring, and—according to its proponents—essential for maintaining American leadership in deep space. Both perspectives deserve fair hearing. What follows is the unvarnished account of how much this actually costs, how close the program has come to cancellation, and what must happen for it to survive and succeed.

NASA's total Artemis spending through fiscal 2025, per the Inspector General, breaks down approximately as:

- SLS development and first four launches: $28 billion

- Orion development and first four missions: $29 billion

- Exploration Ground Systems: $5.5 billion

- Gateway, landers, suits, and supporting systems: Remainder of ~$93 billion total

Per-launch costs for SLS Block 1 and Orion combined approach $4.1 billion when development costs are amortized across early missions. As production matures and fixed costs are spread over more launches, this figure should decline, though NASA has not published target costs for mature operations.

For comparison, SpaceX's Falcon Heavy launches for approximately $150 million in expendable mode, though it cannot currently deliver crew to lunar orbit. Starship aims for dramatically lower costs through full reusability, but has not yet reached orbit and remains years from human-rating certification. The cost comparison is thus between an operational system (SLS/Orion) and developmental ones (Starship, New Glenn) with unproven capabilities.

The Price Tag

NASA's Office of Inspector General estimated in 2023 that the Artemis program would cost approximately $93 billion through fiscal year 2025, with ongoing expenses continuing beyond that date. Breaking that down: Space Launch System development and first four launches account for roughly $28 billion. Orion development and the same four missions add another $29 billion. Ground systems, including the mobile launcher and renovations to Kennedy Space Center facilities, contribute approximately $5.5 billion. The remainder funds Gateway, lunar landers, spacesuits, and program management. When development costs are amortized across early missions, each Space Launch System and Orion launch costs approximately $4.1 billion.

The comparison with commercial alternatives is stark and unavoidable. SpaceX's Falcon Heavy, currently the most powerful operational rocket, launches for approximately $150 million in expendable configuration—roughly one twenty-seventh the cost of an SLS/Orion launch. The company's Starship, once operational, aims for dramatically lower costs through full reusability. Blue Origin is developing New Glenn with similar cost-reduction goals. China is developing the Long March 9. If commercial vehicles can lift comparable payloads for a fraction of the price, the burden of justification falls on NASA to explain why heritage hardware remains necessary.

The justification, though it fails to silence critics, rests on timing and proven technology. When Artemis II takes flight, the emerging commercial heavy-lift vehicles remain unproven for human deep-space exploration. Neither Starship's orbital capability nor New Glenn's launch record will have been established. China's Long March 9 remains theoretical. NASA's position is straightforward: the agency required operational heavy-lift capacity in the current decade rather than waiting for alternatives to mature in the 2030s. The Space Launch System, built on hardware that logged three decades of Shuttle flights, offered that proven capability when the program needed it.

Second, human-rating a launch vehicle imposes requirements that go beyond raw performance. The rocket must demonstrate extremely high reliability, provide abort modes throughout ascent, undergo exhaustive testing, and incorporate redundancy in critical systems. NASA's design margins for crewed vehicles are more conservative than those typically used by commercial operators willing to accept higher risk in exchange for rapid iteration. Whether that conservatism is justified or excessive depends on one's tolerance for risk, but the agency's post-Challenger and post-Columbia culture prioritizes crew safety even when that drives up costs.

Third, political and economic factors shape these decisions as much as engineering considerations. The Space Launch System sustains thousands of high-skill jobs across the United States. Manufacturing facilities in Louisiana, Alabama, Mississippi, Utah, and Florida depend on the program. Congressional representatives from those states have consistently championed SLS funding, and the 2025 One Big Beautiful Bill Act injected nearly $10 billion into NASA's budget with language mandating continued development beyond the first lunar landing. Critics call this pork-barrel spending; supporters argue it maintains an industrial base that would take decades to rebuild if allowed to collapse.

NASA's strategy attempts to balance government capability with commercial innovation. The agency uses SLS for crew while contracting with SpaceX for the Artemis III lunar lander, with Blue Origin also developing a lander for future missions. Axiom Space is building next-generation spacesuits. United Launch Alliance and other commercial providers deliver cargo and smaller payloads. This hybrid model aims to leverage government-funded assured access for crew while stimulating commercial competition for other services. Whether the balance is right remains a subject of legitimate debate, but the approach recognizes that no single model—pure government or pure commercial—optimally serves all mission requirements.

What's undeniable is that Artemis is expensive by almost any measure. The $4 billion per launch figure exceeds the entire annual budget of many federal agencies. For that same money, NASA could fund multiple robotic missions to Mars, Jupiter, or Saturn that would return transformative science. It could invest in telescope observatories that peer back to the universe's first moments. It could accelerate Earth science missions addressing climate change. These are real opportunity costs that cannot be dismissed as irrelevant.

Supporters respond that national priorities include more than maximizing scientific return per dollar. They argue that human spaceflight serves purposes that robotic missions cannot: inspiring the next generation, demonstrating technological leadership, building international coalitions, and developing capabilities that could prove strategically important as other nations pursue lunar ambitions. The cost question ultimately reduces to priorities. Is human spaceflight beyond low Earth orbit worth billions per year? Should taxpayers fund a program whose benefits are long-term, diffuse, and partly intangible? Or should those resources address more immediate terrestrial needs? These are questions of values, not just economics, and readers approaching them honestly may reach different conclusions.

Political Survival

Artemis has survived at least three serious attempts to scale it back or terminate it entirely since its inception. Each time, the program adapted, found new champions, or benefited from timing that made cancellation politically difficult. Understanding these close calls reveals both the program's resilience and its fragility.

The first major threat came during the Trump administration's budget proposals. In 2019, the President's budget request included a stunning proposal to cut NASA's science budget by roughly fifty percent while increasing human spaceflight funding. The move was framed as prioritizing exploration over science, but it triggered fierce opposition from the scientific community and from congressional appropriators who valued programs like the James Webb Space Telescope and planetary missions. The budget proposal went nowhere—Congress restored the science cuts and maintained Artemis funding—but it exposed tensions between different NASA constituencies and raised questions about whether Artemis would come at the expense of other programs.

The second near-death experience came after Artemis I's successful flight in late 2022. With the uncrewed test complete and Artemis II years away, some budget hawks in Congress proposed cutting funding beyond the initial lunar landing, arguing that a single crewed landing would satisfy the program's goals and that sustaining Gateway and additional missions was too expensive. The argument gained traction among fiscal conservatives concerned about deficit spending. If momentum had shifted decisively in that direction, the program might have been reduced to a flags-and-footprints repeat of Apollo—one landing, maybe two, then cancellation.

What saved Artemis was a combination of bipartisan support and strategic timing. Legislators from states with NASA facilities and contractors formed a powerful coalition defending the program. International partners, having invested billions in Gateway modules and Orion's service module, applied diplomatic pressure to continue. And crucially, China's lunar ambitions became more visible and concrete, making cancellation of Artemis look like ceding leadership in space at exactly the moment when competition was intensifying. The geopolitical argument proved more persuasive than the fiscal one.

The third threat—and the closest call—came during the contentious 2025 budget negotiations. Artemis faced possible delays or restructuring until the One Big Beautiful Bill Act injected nearly $10 billion into NASA's budget with explicit language supporting continuation beyond Artemis III. That legislative package, controversial for many reasons beyond space exploration, nonetheless stabilized Artemis funding through the mid-2020s and enabled planning for sustained lunar operations. The bill passed with unusual bipartisan margins, suggesting that space exploration retains political support even in polarized times, though that consensus remains fragile.

Each survival episode required compromise. Schedules slipped to accommodate budget realities. Certain Gateway modules were delayed or descoped. The lunar terrain vehicle and other surface infrastructure moved right on the timeline. Critics argue these compromises prove the architecture is unaffordable; supporters respond that adapting to fiscal constraints while maintaining core objectives demonstrates pragmatism. Both perspectives have merit. What's clear is that Artemis exists in a state of perpetual political negotiation, never secure enough to plan confidently beyond a few years, never endangered enough to face outright cancellation.

The fragility matters. Long-term programs require long-term commitments. International partners hesitate to invest in systems if they suspect the program might be canceled before those systems fly. Contractors struggle to maintain workforces and production lines when funding arrives in sporadic bursts rather than steady streams. Engineers designing spacecraft for missions a decade away must guess whether political support will last that long. This uncertainty adds cost—workforce disruptions, stop-start manufacturing, redesigns to accommodate budget cuts—that compounds the fiscal challenges the program already faces.

Yet the program has survived, and that survival reflects something real: a political consensus, however fragile, that returning humans to the Moon matters. That consensus spans conservative legislators who value military and strategic space capabilities, progressives who support science and international cooperation, and moderates who see economic benefits in NASA jobs. It includes international partners who've invested billions and expect the United States to follow through. It draws strength from public enthusiasm that peaks around major milestones like launches and landings. Whether this coalition can sustain Artemis through the 2030s and beyond remains uncertain, but the program's resilience through repeated near-death experiences suggests it has deeper roots than critics often acknowledge.

Oversight and Valid Concerns

NASA operates under constant scrutiny from multiple oversight bodies, and their assessments of Artemis have been consistently skeptical. The Government Accountability Office, the NASA Inspector General, and Congressional Research Service have published dozens of reports questioning cost estimates, schedule projections, technical risks, and program management. Dismissing these concerns as bureaucratic nitpicking misses the point: many of the criticisms are well-founded and serve important accountability functions.

The GAO has repeatedly flagged the Human Landing System as the program's highest technical risk. SpaceX's Starship HLS requires technologies that have never been demonstrated at operational scale, most notably cryogenic propellant transfer in orbit. The concept involves launching multiple tanker flights—perhaps seven, perhaps fifteen, depending on whose estimates you trust—to refuel a depot in orbit, which then transfers propellant to the lunar lander. Each transfer involves docking, thermal management to prevent boil-off, and fluid dynamics in microgravity. If any step fails or proves more difficult than expected, the entire Artemis III schedule collapses.

NASA and SpaceX are conducting ground tests and planning an orbital demonstration for 2026, but these tests will occur very close to the planned Artemis III launch date, leaving little margin for discovering problems and fixing them. The GAO argues—reasonably—

that this compressed timeline increases risk and that NASA should have required earlier demonstrations or selected a less ambitious lander design. NASA's response is that SpaceX's approach offers capabilities that more conventional designs cannot match and that the risk is acceptable given the strategic importance of landing before China's target date. Both positions have merit; the disagreement reflects different risk tolerances rather than one side being clearly wrong.

The Inspector General has focused extensively on cost growth and schedule delays. Nearly every major Artemis component has exceeded original budget estimates and missed original timelines. The mobile launcher, originally budgeted at $234 million, ended up costing more than $1 billion. Orion's development stretched from initial projections of six years to more than fifteen. Ground systems renovations ran billions over budget. The IG argues these overruns reflect poor management, inadequate oversight, and contractors exploiting cost-plus contracts that provide little incentive to control expenses.

NASA's response acknowledges the delays and cost growth but attributes much of it to technical complexity, changing requirements, and insufficient initial funding. When Congress appropriates less than requested, programs stretch out, which paradoxically increases total cost as teams must be maintained longer. When test programs discover problems—like the Orion heat shield issue—addressing them takes time and money. When requirements change midstream—such as when

Gateway's orbit was modified from the original plan—already-designed hardware must be reworked. The IG's critiques highlight real failures, but the solutions are not always obvious: tighter management can help, but it cannot eliminate the fundamental uncertainties of developing cutting-edge technology.

Congressional Research Service reports have questioned the sustainability of the architecture. If each launch costs $4 billion, how many can NASA afford per year? Current budgets support roughly one launch annually, maybe two. That pace allows for a crewed lunar mission every year or two, enough for the initial exploration phase but inadequate for building a true lunar economy. Sustaining Gateway, supporting surface operations, maintaining ground infrastructure, and funding Mars development simultaneously would require budgets NASA is unlikely to receive. The CRS argues this creates a "sandbox problem"—NASA is building impressive castles in the sand, but the tide of fiscal reality may wash them away before they're finished.

These oversight concerns deserve serious consideration. They identify genuine vulnerabilities that could undermine the program's success or sustainability. Ignoring them because they're inconvenient or because you support Artemis does not serve the program's long-term interests. Better to acknowledge the problems, work to address them, and build political support for solutions than to pretend everything is fine when oversight bodies clearly document that it isn't.

At the same time, oversight reports have limitations. They often emphasize risks and problems while giving less weight to progress and successes. They compare actual performance against idealized timelines that may have been unrealistic from the start. They sometimes recommend risk mitigation strategies that would add years to schedules—exactly the outcome they critique when it happens. And they operate with perfect hindsight, identifying problems that seem obvious after the fact but were genuinely uncertain when decisions were made. The value of oversight is holding programs accountable; the limitation is that accountability reports can create a perception of dysfunction even when programs are navigating genuine technical challenges reasonably well.

Economic and Strategic Value

The case for Artemis rests on several pillars that extend beyond pure science or exploration. The economic argument begins with jobs—NASA spending on Artemis supports roughly 70,000 direct positions across the aerospace industry, with indirect employment in suppliers and local communities adding tens of thousands more. These are high-skill, high-wage positions that anchor regional economies and keep advanced manufacturing capability on American soil.

Beyond direct employment, space technology generates spinoffs that benefit other sectors. Apollo gave us integrated circuits, water purification systems, and materials that found applications far beyond spaceflight. Space Shuttle technology contributed to medical imaging and aircraft design. Artemis will drive advances in life support, autonomous operations, radiation shielding, and additive manufacturing that have terrestrial applications. These spinoffs are difficult to quantify in advance, but historical patterns suggest substantial long-term returns on space investment.

The strategic argument focuses on leadership and competition. Space has become a domain of national power, and nations that excel in space tend to shape the rules and norms that govern its use. China's rapid progress in lunar exploration represents a direct challenge to American preeminence. If China establishes a presence at the lunar south pole first, it may set precedents for resource extraction and territorial claims that other nations will find difficult to contest. The Artemis Accords represent an American-led framework for lunar governance, but frameworks only matter if the nations advancing them have actual operational presence.

The inspirational dimension is harder to quantify but may be most important. Human spaceflight captures imaginations in ways that robotic missions typically do not. Students who watch Artemis II are statistically more likely to pursue careers in science, technology, engineering, and mathematics. The diversity of the crew—the first woman, first person of color, first Canadian beyond low orbit—sends powerful signals about who belongs in these fields. The long-term benefits to the STEM workforce pipeline may justify a significant fraction of the program's cost even if no other returns materialize.

Critics respond that these arguments justify some human spaceflight program but not necessarily Artemis at its current cost. Perhaps a smaller, less ambitious program would deliver similar benefits more efficiently. Perhaps international cooperation could share costs and reduce the American burden. Perhaps robotic missions plus commercial space tourism would provide sufficient inspiration without the expense of government-funded lunar expeditions. These are fair critiques that deserve consideration.

What's clear is that millions of Americans—and millions more around the world—find value in the endeavor beyond what cost-benefit spreadsheets capture. When Artemis II launches, tens of millions will watch. Schools will pause classes. Families will gather around screens. People who cannot articulate why space exploration matters will nonetheless feel that it does, that pushing beyond our planet says something important about human ambition and possibility. Whether that justifies the expense is ultimately a political question, not a technical one. But the persistence of public support despite budgetary pressures suggests the answer for many people remains yes.

What Success Looks Like

Artemis stands at a crossroads. Artemis II will demonstrate that the core systems work with crew aboard. Artemis III will attempt the first lunar landing in more than fifty years. But what comes after determines whether the program justifies its cost or joins Constellation as another exploration architecture that flew a few missions and then faded.

Success requires sustaining political and fiscal support through the 2030s and beyond. That means demonstrating value—conducting science that matters, advancing technology that transfers to other domains, maintaining international partnerships that distribute costs and benefits broadly. It means managing costs more effectively than the program has to date, finding efficiencies that make each launch more affordable, and transitioning toward commercial services where they can deliver capabilities reliably and cheaply. It means adapting when systems don't work as planned, as the heat shield issue demonstrated, rather than letting problems cascade into program-threatening crises.

It also requires building a constituency beyond aerospace workers and space enthusiasts. Gateway must serve as more than a waystation—it needs to produce science and enable operations that justify its cost. Lunar surface missions must yield discoveries that capture broader public interest, not just geologists and planetary scientists. The path to Mars must remain visible and credible, ensuring that Artemis is understood as a means to an even more ambitious end rather than as an end in itself.

The program's fragility is not a weakness to hide but a reality to acknowledge and address. No major exploration initiative survives on autopilot. Apollo succeeded because it had a clear deadline, powerful political backing, and focused public attention. When those ingredients faded, the program ended despite having hardware built for additional missions. Artemis has no firm deadline beyond racing China, no single champion as powerful as President Kennedy, and faces public attention that peaks during launches but wanes between them. Whether it succeeds long-term depends on whether supporters can build durable political coalitions, deliver results that justify continued investment, and adapt to fiscal realities without abandoning core objectives.

Honest assessment demands acknowledging both the program's genuine achievements and its serious vulnerabilities. Artemis has survived repeated attempts to cancel it, built hardware that works, and assembled an international coalition supporting sustained lunar exploration. Those are real accomplishments. It has also experienced cost overruns, schedule delays, and technical challenges that vindicate many oversight criticisms. Those are real problems. Both can be true simultaneously.

The hardest truth is that expensive, ambitious programs always generate legitimate criticism, and dismissing that criticism as obstructionist or uninformed does not serve the program's interests. Better to engage critics seriously, acknowledge valid concerns, work to address them, and make the strongest possible case for why the endeavor remains worthwhile. Artemis will be judged ultimately not by whether it faced criticism—all programs do—but by whether it delivered value commensurate with its cost and whether it built the foundation for sustained exploration that its advocates promise. That judgment still lies ahead, shaped by the missions now beginning and the political will that sustains or abandons them in the years to come.

Rendering of future base at Shackleton Crater.
(MS Designer)

CHAPTER EIGHT

The Architecture

Artemis II validates that the core systems work—that the Space Launch System and Orion can carry crews safely to lunar distance and back, that manual flight capability enables the rendezvous operations Gateway will require, that the lofted entry profile protects returning astronauts through an inferno never before experienced with this hardware. But validation is only the beginning. The transportation pipeline Artemis II proves enables everything that follows: landing missions to the lunar south pole, an orbital outpost that serves as staging point and research platform, surface infrastructure for long-duration stays, and eventually the pathway to Mars.

Think of the architecture as a roadmap, not a comprehensive blueprint—the details will evolve as missions fly and technology matures, as budgets shift and international partners contribute, as commercial capabilities prove themselves or fall short. But understanding what comes next helps contextualize why Artemis II's validations matter so much. When Victor Glover manually

Opposite: Rendering of future base at Shakelton Crater.
(MS Designer)

flies Orion during the proximity operations demonstration, he's not just testing the spacecraft—he's certifying the capability that future landing missions absolutely require. When Christina Koch monitors life support for ten days, she's validating systems that must work flawlessly before crews attempt weeks-long surface stays. When the crew returns through the lofted entry, they're proving that all subsequent Artemis crews can safely come home.

The architecture isn't static. NASA is building Gateway while commercial partners develop landers. International contributions arrive on different schedules. Budget realities will force prioritization and delays. Some elements described here will fly largely as planned; others will be descoped, redesigned, or canceled. But the vision—sustained human presence on and around the Moon, building toward Mars—remains the north star that guides development. What follows describes where Artemis is heading if the political will and fiscal resources continue.

Artemis III: The Landing

If Artemis II succeeds, Artemis III becomes the lunar landing NASA has prepared for since the program began. Current planning targets late 2027 or 2028, though that schedule depends on SpaceX completing the Human Landing System and demonstrating orbital refueling. The mission architecture is complex, involving multiple launches and rendezvous operations in lunar orbit.

The Artemis III crew—four astronauts again, though possibly with different assignments than Artemis II— will launch aboard SLS and Orion just as Wiseman's crew does. But instead of swinging around the Moon and returning directly, they'll enter a near-rectilinear halo orbit, or NRHO, where Gateway will eventually reside. There they'll rendezvous with SpaceX's Starship Human Landing System, already waiting in lunar orbit after its own journey from Earth.

The Starship HLS is enormous—roughly 50 meters tall, capable of carrying large payloads and providing spacious accommodations compared to Apollo's cramped Lunar Module. It launched separately, probably weeks or months before the crew, and spent time in low Earth orbit receiving propellant from multiple tanker flights. Think of it like this: Starship launches to orbit nearly empty. Then six or eight or perhaps a dozen tanker Starships launch, each carrying propellant that they transfer to the HLS. Once fully fueled, the HLS fires its engines to reach lunar orbit and waits for the crew to arrive.

Two astronauts—the commander and one mission specialist—will transfer from Orion to the Starship HLS via docking adapters, leaving two crew members aboard Orion to maintain the spacecraft and serve as backup if the landing team encounters problems. The manual flight capability that Glover validates during Artemis II's proximity operations demonstration becomes essential here—automated systems handle the approach, but astronauts must be able to assume control if sensors disagree or unexpected situations arise. The training from Artemis II directly informs how landing crews prepare for this critical rendezvous.

The two landing astronauts will descend to the lunar south pole, targeting sites near permanently shadowed craters where water ice is believed to exist. Unlike Apollo's landings on relatively flat maria near the equator, these sites feature rugged terrain with steep slopes and boulders that demand precise navigation. The Starship HLS will use terrain-relative navigation—comparing camera images against stored maps to identify safe landing zones autonomously. Apollo astronauts manually flew the final approach; Artemis III astronauts will monitor and supervise but will likely let automation handle most of the descent unless problems arise.

On the surface, the crew will spend roughly a week— far longer than Apollo's maximum of three days. They'll conduct multiple moonwalks using Axiom Space's next-generation suits, which provide greater mobility than the Apollo A7L suits and can sustain longer durations outside. They'll deploy scientific instruments, collect samples from permanently shadowed regions, and test technologies for extracting water ice. They'll demonstrate that humans can live and work effectively in the challenging south pole environment, where sunlight arrives at extreme low angles and crater floors remain in permanent darkness at temperatures cold enough to trap volatiles for billions of years.

When the surface mission concludes, the landing crew ascends in Starship HLS, returns to lunar orbit, and docks again with Orion. They transfer back, and Orion fires its European Service Module engines to leave lunar orbit and return to Earth. The entire mission duration might stretch to three weeks—longer than any Apollo flight, testing life support and crew endurance at scales not attempted since Skylab. The experience Koch gains monitoring environmental systems during Artemis II's ten days provides crucial baseline data that informs how engineers prepare for Artemis III's extended mission.

The technical challenges are formidable. Orbital refueling has never been demonstrated at operational scale. Starship must prove it can land reliably on the Moon—SpaceX plans uncrewed demonstration missions first, but those must succeed before NASA commits a crew. The permanently shadowed craters present navigation difficulties and thermal extremes. And any significant delay in Starship development pushes the entire mission timeline right, potentially opening a window for China to land taikonauts first.

But if it works—if Artemis III lands successfully and returns safely—the mission will mark the first time humans have stood on the Moon since Eugene Cernan climbed back into the Apollo 17 Lunar Module in December 1972. It will make real the promise that Artemis is more than Apollo 2.0, delivering capabilities and durations that the 1960s program never achieved.

Rendering of Space Station Gateway.
(MS Designer)

Gateway: The Lunar Outpost

Gateway is NASA's small space station planned for near-rectilinear halo orbit around the Moon. Unlike the International Space Station, which circles Earth every 90 minutes in low orbit, Gateway will follow a seven-day elliptical path that swoops close to the Moon's north pole and then swings high above the south pole. This unusual orbit is gravitationally stable, requiring minimal propellant to maintain, and provides access to most of the lunar surface for landing missions.

The station will be much smaller than ISS—think of it as a small apartment rather than a sprawling complex. Initial modules include the Power and Propulsion Element, built by Maxar Technologies, which provides solar electric propulsion and 60 kilowatts of power. The Habitation and Logistics Outpost, built by Northrop Grumman, offers living quarters for crews during Gateway missions. Europe and Japan are contributing additional modules: the International Habitation Module will expand living space and provide life support, while a logistics module will enable resupply and refueling.

Canada's contribution—the Canadarm3 robotic system—will handle external maintenance, capture visiting vehicles, and support science operations. The system builds on decades of Canadian expertise with robotic arms that served the Space Shuttle and ISS, extending that capability to deep space where communication delays and harsh radiation require greater autonomy. The communications systems tested during Artemis II's lunar blackouts—learning how crews

Gateway's orbit is a near-rectilinear halo orbit, abbreviated NRHO, which is gravitationally stable and requires minimal propellant to maintain. The orbit takes seven days to complete, with its lowest point—the perilune—passing about 3,000 kilometers above the Moon's north pole and its highest point—the apolune—reaching more than 70,000 kilometers above the south pole.

The orbit's advantages include stability (it doesn't decay or require frequent correction burns), communications (most of the time Gateway can see Earth for radio contact), and access (landers can reach most of the lunar surface from this orbit without excessive fuel requirements). The disadvantage is low gravitational force—objects in NRHO experience extremely weak "gravity" that's less than one-tenth of what the ISS experiences. This means procedures for moving equipment, managing fluids, and handling emergencies must account for near-zero-G conditions.

The orbit was chosen after extensive analysis comparing alternatives including low lunar orbit (which requires more propellant to maintain and blocks communications when passing behind the Moon) and Lagrange points (which provide stability but complicate landing operations). NRHO represents the compromise that best satisfies Gateway's multiple requirements.

function when separated from Earth's continuous contact—directly inform how Gateway operations are designed, ensuring that station crews can work effectively even when passing behind the Moon interrupts ground communications.

Gateway serves multiple purposes. It's a staging point where crews transferring from Orion to lunar landers can prepare and rest. It's a logistics hub where supplies can be pre-positioned before surface missions. It's a research platform conducting experiments in deep space that couldn't be done in low Earth orbit—studying solar radiation, testing long-duration life support, and practicing operations far from Earth. And it's an international facility that deepens partnerships beyond what ISS achieved, with modules contributed by multiple nations and crew rotations including astronauts from the Artemis Accords signatories.

Critics question whether Gateway is necessary. SpaceX's direct-landing architecture for Starship might bypass the station entirely—why stop at Gateway if you can launch from Earth, land on the Moon, and return without intermediate staging? The answer involves risk management and flexibility. Gateway provides a safe haven if problems arise during surface missions. It enables "split missions" where cargo and crew launch separately, reducing the mass any single rocket must lift. It allows reusable lunar landers to refuel and fly multiple missions rather than being expended each time. And it serves as a testbed for Mars transit habitats, where crews will spend months in deep space without the option to abort back to Earth.

Assembly of Gateway will begin after Artemis III,

with modules launched on commercial rockets and integrated in orbit. Early missions will be uncrewed, with systems checked out remotely. Later missions will bring crews for weeks at a time, gradually building operational experience before committing to the longer stays required for Mars preparation. The station's small size means it won't support continuous habitation like ISS does—crews will come and go, sometimes leaving Gateway uncrewed for months between visits. This operational model reduces costs but requires robust automation and fault-tolerance since ground controllers can't always respond quickly to anomalies.

Gateway's orbit has another advantage: communications satellites placed in cislunar space can relay signals between Earth and the lunar far side, enabling missions to permanently shadowed craters that would otherwise lose contact whenever their sites rotated out of view from Earth. This infrastructure becomes increasingly important as lunar operations expand and multiple missions operate simultaneously. Gateway coordinates traffic, manages communications, and ensures that visiting vehicles don't interfere with each other—air traffic control for the Moon.

Whether Gateway justifies its cost remains a subject of debate. Supporters see it as essential infrastructure that enables sustained exploration. Critics see it as an expensive intermediate step that could be bypassed by more capable vehicles. The station's value will be judged ultimately by whether it enables missions that wouldn't otherwise be possible and whether it proves its worth as a Mars testbed. Those judgments lie years ahead, shaped by missions that haven't yet flown.

Rendering of Lunar Terrain Vehicle, or LTV.
(MS Designer)

Surface Systems: Living and Working on the Moon

Landing once is impressive; living and working on the Moon requires infrastructure. NASA and its partners are developing systems that transform the lunar surface from a destination you visit briefly into a place where crews can stay for weeks or eventually months at a time.

The Lunar Terrain Vehicle, or LTV, is essentially a car for the Moon. Unlike Apollo's Lunar Roving Vehicle, which was used for a few hours across three missions then left behind, the LTV is designed for long-term use, capable of driving autonomously when uncrewed and transporting astronauts during surface operations. Multiple companies are competing to build it—each proposing different designs, but all meeting requirements for range, cargo capacity, and reliability. The winning design will support missions across the south pole region, allowing crews to travel dozens of kilometers from their landing site to collect samples, deploy instruments, and scout future base locations.

Japan is developing a pressurized rover called the Lunar Cruiser—a camper-van-sized vehicle where astronauts can live for days at a time without wearing spacesuits. The Cruiser extends exploration range dramatically. Instead of returning to the lander each night, crews could drive for a week, conducting geology along the way and sleeping in the rover's cabin. It's an audacious concept that requires sophisticated life support, reliable power, and navigation systems capable of operating far from Earth-based controllers. If successful, it could enable missions to permanently shadowed craters and mountain peaks that lie beyond the range of daily excursions from landing sites.

Habitats for long-duration stays are under study by multiple contractors. These structures must withstand temperature extremes, provide radiation shielding, maintain air pressure despite micrometeorite impacts, and generate power through the fourteen-day lunar night. Some concepts use inflatable modules that pack efficiently for launch but expand once deployed. Others propose rigid structures built from lunar regolith using 3D printing—essentially manufacturing bricks from Moon dirt and assembling them into protective shells. The life-support monitoring experience Koch gains during Artemis II feeds directly into habitat design—understanding how systems respond to crew metabolic loads in confined spaces under deep-space radiation informs the requirements for surface habitats where those challenges persist for weeks.

Power systems are critical. Solar arrays work during the two-week lunar day, but surface temperatures plunge during the night and solar panels generate nothing. Batteries can bridge short gaps, but weeks of darkness require alternative power. NASA is developing fission surface power—small nuclear reactors that generate steady electricity regardless of whether the sun is up. A 10-kilowatt reactor operating continuously could support significant operations, providing power for habitats, experiments, mining equipment, and propellant production. The technology carries political sensitivities—launching nuclear material raises safety concerns, and deploying reactors on the Moon raises environmental and treaty questions—but the energy density of fission may be necessary for sustained operations.

In-situ resource utilization, or ISRU, aims to extract water ice from permanently shadowed craters and process it into oxygen and hydrogen. Oxygen supports breathing and replenishes life support systems; hydrogen serves as propellant. The concept dramatically reduces the mass that must be launched from Earth. Instead of carrying all consumables for a mission, crews could manufacture them on-site, living off the land in a sense. Early demonstrations will be small—processing a few kilograms of ice as proof of concept. But the long-term vision involves industrial-scale operations where lunar resources support not just exploration but economic activity—refueling spacecraft headed to Mars, manufacturing construction materials, perhaps eventually exporting resources to Earth orbit where they're more valuable than lifting equivalent mass from Earth's gravity well.

Communications infrastructure expands alongside operational capabilities. Relay satellites placed in stable orbits around the Moon provide continuous coverage for surface operations, enabling crews to stay in contact even when their landing site faces away from Earth. These satellites also support navigation, providing GPS-like positioning services accurate to meters. Optical communications using lasers increase bandwidth dramatically over traditional radio, enabling high-definition video and large data transfers that make remote science collaboration more effective.

All these systems—rovers, habitats, power, ISRU, communications—must work together as an integrated architecture. A habitat without reliable power is useless. A rover without communications is dangerous. ISRU without power and habitats to support the crews operating the equipment doesn't accomplish anything. NASA describes this as building the "Artemis Base Camp"—not a single structure but a distributed network of capabilities that collectively enable sustained presence.

The timeline for deploying these systems stretches across the late 2020s and into the 2030s. Early missions will demonstrate individual technologies. Later missions will integrate them into operational systems. The lunar south pole will gradually transform from pristine wilderness into a work site, with landing pads, rovers leaving tracks, habitats casting shadows, and crews conducting the first sustained human operations on another world. Whether that transformation happens depends on political will, fiscal resources, and technical success on missions beginning with Artemis II's validation of the transportation systems that make everything else possible.

The Mars Connection

Every Artemis system—the SLS rocket, Orion capsule, Gateway station, landers, habitats, ISRU equipment— is designed with Mars in mind. The Moon serves as a proving ground where technologies can be tested under conditions that approximate Mars without committing to the eight-month journey that makes rescue impossible.

Deep-space radiation is the first challenge. Beyond Earth's protective magnetosphere, cosmic rays and solar particles bombard spacecraft continuously. Gateway's orbit spends most of its time well outside that protection, exposing crews to radiation levels similar to what they'd experience traveling to Mars. The AVATAR experiment flying on Artemis II—testing bone marrow cells in deep-space conditions—represents the first step in understanding how long-duration exposure affects human health. Gateway missions lasting weeks to months will gather more data, informing shielding strategies, medical countermeasures, and acceptable exposure limits for Mars crews.

Life-support systems must operate reliably for years, not weeks. The Environmental Control system on Artemis II processes carbon dioxide and recycles water for ten days; Mars transit habitats must do the same for 500 days or more. Gateway provides the testbed for proving these systems at scale. Engineers need to understand failure modes, maintenance cycles, how to repair equipment when spare parts are limited, and how much margin exists before degradation threatens crew safety. Every day Gateway operates with crew aboard generates data that makes Mars life support more reliable.

Autonomous operations become essential at Mars distances. When communication delays stretch to twenty minutes round-trip, crews cannot wait for ground controllers to diagnose problems or provide instructions. The autonomous operations tested during Artemis II's lunar blackouts—when Orion passes behind the Moon and crews operate independently for up to an hour—scale to Mars where independence must be sustained for months. Gateway crews will practice making decisions without Earth's constant oversight, building confidence that Mars crews can function effectively when truly alone.

ISRU on the Moon directly translates to Mars. If you can extract water from lunar regolith and split it into oxygen and hydrogen, the same techniques work with Martian ice and atmospheric processing. If you can generate oxygen for breathing and propellant for rockets using local resources, you've eliminated the need to carry those consumables across interplanetary space—a mass savings that might make Mars missions affordable where launching everything from Earth would be prohibitively expensive. The Moon lets NASA demonstrate these capabilities with margin for error; if ISRU fails on the Moon, crews can still return safely. Failure on Mars could be catastrophic.

Propellant production is particularly critical. A Mars

mission might require 30 to 40 metric tons of propellant for the return journey. Launching that much from Earth is possible but expensive and requires massive rockets. Producing it on Mars using water extracted from ice deposits and carbon dioxide from the atmosphere reduces the mass that must be carried, potentially enabling missions that would otherwise be unaffordable. But the technology must be proven before committing crews. Demonstrating it on the Moon—where the chemistry differs but the principles are similar—builds confidence.

Gateway's orbit serves another purpose: it's a prototype for deep-space habitats that will eventually carry crews to Mars. The station's small size, autonomous operations, and modest crew complement mirror what Mars transit vehicles might look like. Engineers learn how to design for long-term habitation in deep space where resupply is difficult and rescue is impossible. They discover what breaks, what works better than expected, and what crews actually need versus what planners thought they needed. That knowledge shapes Mars vehicle design in ways that ground testing cannot replicate.

The Artemis timeline deliberately paces Moon missions to build capability before attempting Mars. Early missions test transportation and short-duration stays. Middle missions establish Gateway and demonstrate extended surface operations. Later missions operate the full architecture—reusable landers, functioning habitats, active ISRU, continuous presence. Each phase validates systems that Mars will require, and each phase identifies problems that can be fixed before committing to the Red Planet.

NASA targets the late 2030s or early 2040s for the first crewed Mars mission, though that timeline depends on budgets, political support, and technical progress. The Moon doesn't just enable Mars—it makes Mars survivable. By the time humans travel to Mars, they'll be using life support proven on Gateway, ISRU techniques demonstrated at lunar base camps, autonomous operations refined during years of Moon missions, and spacecraft architectures tested in deep space. None of that experience can be gained in low Earth orbit, where rescue is hours away and resupply launches every few weeks. The Moon is close enough to be risky but far enough to teach the hard lessons.

Critics argue that Mars could be reached faster if NASA abandoned the Moon and focused exclusively on the Red Planet. They point to early studies suggesting Mars missions could launch in the 2030s if properly funded. They note that the Moon and Mars present different challenges—what works on the Moon may not translate directly—and that lunar detours consume budgets better spent on Mars hardware. These are serious arguments that deserve consideration.

Supporters counter that attempting Mars without extensive deep-space experience invites catastrophic failure. They note that every crewed Mars architecture requires systems—life support, ISRU, autonomous operations—that must be proven somewhere before committing crews to a mission where return takes years. They argue that the Moon provides that proving ground at acceptable risk, and that the international partnerships and public support Artemis builds make Mars politically sustainable where a unilateral American Mars shot might collapse when budgets tighten or political winds shift.

Both perspectives have merit. What's clear is that NASA has committed to the Moon-first pathway, for better or worse. Whether that strategy proves wise will be judged by whether it delivers crews to Mars in the 2040s with systems that work and budgets that Congress sustains. If Artemis succeeds and Mars follows, the lunar detour will be remembered as essential preparation. If Artemis stalls and Mars recedes into the 2050s or beyond, critics will argue that focusing on the Moon squandered the window when Mars was achievable. The answer won't be known for decades, but the pathway is set, beginning with Artemis II's validation that the core systems work.

Earthrise - December 2017
(Photo courtesy of N.A.S.A)

CHAPTER NINE

Significance of Artemis

You've followed the technical details: the 8.8 million pounds of thrust that will accelerate four astronauts away from Earth, the proximity operations demonstration that validates manual flight for future Gateway docking, the lofted entry profile born from engineering necessity that protects the crew through temperatures reaching 2,760 degrees Celsius. You understand the cost—$4.1 billion per launch—and the political fragility that makes every success essential for program survival. You know the crew: Wiseman, Glover, Koch, Hansen, each bringing skills and perspective the mission requires.

Now the question becomes: why does any of this matter? What's at stake beyond hardware validation and political survival? What does it mean to watch four humans venture farther from Earth than anyone in more than fifty years, knowing they're flying systems that must work or the pathway to sustained exploration closes?

The answers won't satisfy everyone. Those who believe space exploration diverts resources from pressing terrestrial needs will find no argument here that changes that calculation. But for those who see value in pushing beyond our planet, understanding what success means—in multiple dimensions, not just technically—helps frame what you're witnessing when the countdown reaches zero and the vehicle climbs away from Florida.

Defining Success

Artemis II can succeed or fail along multiple dimensions simultaneously, and those dimensions don't map neatly onto each other. Technical success without inspiring anyone matters less than it should. Political success without technical validation is hollow. Understanding these different measures helps assess the mission honestly rather than through simplistic pass/fail frameworks.

Technical success means the crew returns safely with systems validated for operational use. The lofted entry profile works as modeled. Life support maintains the crew for ten days without significant anomalies. Manual flight capability proves reliable enough to certify for Gateway docking. The European Service Module performs as expected. Data collected during the mission informs Artemis III preparations and reduces uncertainty about how hardware behaves with crew aboard. If Artemis II achieves this—safe return with validations complete—NASA can proceed confidently toward lunar landing.

But technical success alone doesn't guarantee the program continues. Artemis I was technically successful yet budget hawks still proposed cuts. The mission must also deliver symbolic success—demonstrating capability in ways that capture attention and justify continued investment. That means moments that resonate: the launch that shakes Florida and reminds millions that America can still do this, the images from lunar distance that put Earth in perspective, the splashdown that proves calculated engineering choices like the lofted entry were sound. Symbolic success is harder to quantify but politically essential. It builds the public support and congressional buy-in that sustain funding when budgets tighten.

Operational success means proving that the architecture can support sustained exploration, not just one-off demonstrations. This includes validating procedures for multi-day missions, confirming that international hardware like the European Service Module integrates seamlessly, demonstrating that crews can operate autonomously during communication blackouts, and showing that mission control can support deep-space operations effectively. Operational success builds confidence that Artemis isn't fragile—that missions can fly regularly rather than requiring years between attempts to refine everything after each flight reveals problems.

Strategic success involves the geopolitical dimension. If Artemis II launches on schedule and returns safely while China's lunar program experiences delays, the United States demonstrates that it can execute complex missions reliably and maintain leadership in deep space. If the mission strengthens international partnerships and expands the Artemis Accords coalition, it proves that cooperative exploration remains viable. If it generates enthusiasm that drives STEM enrollment and workforce development, it builds the human capital future missions require. Strategic success plays out over years, not days, but Artemis II's execution shapes whether those long-term benefits materialize.

The mission could fail technically while succeeding symbolically—returning safely but discovering that certain systems don't validate as hoped, requiring redesign before proceeding. It could succeed technically while failing politically—executing flawlessly but generating so little public interest that budget support erodes. Or it could succeed across all dimensions, providing the foundation Artemis needs to evolve from experimental program into operational capability. Understanding these different success measures helps assess what happens honestly rather than declaring simple victory or defeat.

The Fifty-Year Question

More than fifty years separate Apollo 17's departure from the Moon in December 1972 and Artemis II's planned flight. That gap is longer than the entire history of powered flight from the Wright Brothers to Apollo. It's longer than the span between the first transcontinental railroad and routine commercial air travel. And it raises an uncomfortable question: if humans reached the Moon in 1969 using 1960s technology, why has it taken until 2026 to go back?

The answer involves politics, economics, and shifting priorities more than technology. Apollo succeeded because it had a clear deadline, powerful political backing, and Cold War urgency that justified extraordinary expense. When those factors faded—the Soviet Union effectively conceded the Moon race, public interest waned, and the Vietnam War and domestic programs consumed budgets—support for Apollo collapsed. NASA had hardware built for additional missions. The Saturn V production line was ready to continue. The expertise existed to fly to the Moon repeatedly. But Congress cut funding, and the capability was abandoned.

The lesson is that technical capability alone doesn't sustain exploration. Political will matters more. Apollo demonstrated that humans could reach the Moon; it didn't prove that humans would stay. The question Artemis confronts is whether this time can be different—whether the program can build the political coalitions, international partnerships, and economic value propositions that sustain funding through administrations and Congresses that may not share the original vision.

The gap also reveals something about human civilization's relationship with exploration. For most of history, exploration expanded continuously. Humans migrated across continents, sailed to distant islands, eventually reached both poles and the deepest ocean trenches. Space seemed to follow that pattern through the 1960s—Sputnik, Gagarin, Apollo—but then progress stopped. Not slowed: stopped. No human has ventured beyond low Earth orbit for half a century. The frontier retreated.

Whether Artemis reverses that retreat or proves to be another brief surge before another long pause depends on what follows. If the program leads to sustained presence—Gateway operational, surface missions routine, Mars preparations advancing—the fifty-year gap will be remembered as a regrettable but temporary pause. If Artemis flies a few missions then ends, historians will note that humanity reached out twice, touched the Moon, and retreated both times. Which future materializes depends on choices made in the years ahead, shaped by missions beginning with Artemis II.

Representation as Foundation, Not Justification

You met the crew's diverse composition in Chapter 1: Victor Glover as the first Black astronaut beyond low Earth orbit, Christina Koch as the first woman to the

Moon, Jeremy Hansen as the first Canadian in deep space. You encountered it again when learning about their training, their roles, and the symbolic weight they carry. Rather than repeat those discussions, it's worth acknowledging what representation accomplishes and what it doesn't.

Representation matters because it shapes who sees themselves in these roles, who pursues the education and training required to become astronauts or engineers or mission controllers. When young people see someone who looks like them achieving extraordinary things, the implicit message is "this path is open to you too." That pipeline effect—more diverse students entering STEM fields, more varied perspectives in engineering teams, broader constituencies supporting exploration—delivers real benefits that compound over time.

But representation alone doesn't justify $4.1 billion expenditures. The mission must deliver technical validation, advance exploration objectives, and build capabilities for sustained presence. Representation amplifies those achievements and expands their impact, but it doesn't substitute for them. The crew succeeds by executing their mission effectively, not by meeting demographic criteria. Their diversity makes the success more meaningful and more widely felt, but the success must be real.

The Artemis Generation

The "Artemis Generation" framing positions the program as a response to resignation—a counternarrative to the belief that great projects are relics of earlier eras when political will and fiscal resources aligned in ways they no longer do. The argument holds that if Artemis succeeds, it demonstrates that complex, ambitious, long-term endeavors remain achievable despite polarization, fiscal constraints, and short attention spans. It shows that nations can still commit to goals that won't be realized within one presidential term or congressional session.

For younger generations who came of age after Apollo, space exploration has meant robots to Mars and repair missions to Hubble—impressive achievements but incremental rather than transformational. The International Space Station circles Earth reliably but invisibly, rarely capturing attention beyond enthusiasts. The narrative has been one of consolidation: doing more with less, optimizing operations, commercializing access to low orbit. Important work, but not the stuff of inspiration.

Artemis offers something different: humans going farther than they've been, doing what hasn't been done, attempting what's uncertain. That narrative resonates with people who suspect that civilization has lost the ability to do hard things. Whether you're worried about climate change, pandemic preparedness, infrastructure decay, or geopolitical competition, the capacity to execute complex projects over decades matters enormously. If we can't build the systems required to return to the Moon, what hope is there for the far harder challenges ahead?

The counterargument is that space exploration diverts attention and resources from problems that matter more immediately. Climate change threatens billions; the Moon threatens no one. Poverty, disease, inequality—these demand solutions now, not eventually. Spending billions to send four astronauts around the Moon while terrestrial challenges go unaddressed reflects skewed priorities. This critique is serious and deserves engagement beyond dismissing it as shortsighted.

The response isn't that space exploration is more important than addressing terrestrial problems—it clearly isn't. Rather, it's that the capabilities developed for space have spillovers that benefit Earth, that inspiration and aspiration matter for cultural vitality, and that budgets large enough to support multiple priorities simultaneously can accommodate both space exploration and other investments if political will exists. NASA's entire budget is roughly 0.5% of federal spending. Artemis consumes a fraction of that. The question isn't whether we can afford space exploration given terrestrial needs; it's whether we choose to prioritize both.

The Artemis Generation framing, if it takes hold, suggests we can. It positions younger people as inheritors of exploration rather than spectators to its decline. It frames the program as something they participate in—by watching, by learning, by entering fields that contribute—rather than something done for them by institutions. And it offers a narrative of possibility at a time when many narratives emphasize constraints.

Whether that framing proves justified depends on whether Artemis delivers sustained exploration rather than another brief surge. If the program leads to Gateway, lunar bases, and eventually Mars, the Artemis Generation will be those who grew up watching it happen and decided to be part of what comes next. If the program stalls after a few missions, the framing will ring hollow. The answer won't be clear until years after Artemis II flies, but the seeds are planted now.

Your Role as Witness

When Artemis II launches, you'll have a choice about what role to play. You can watch passively, noting the event but remaining detached. Or you can engage actively, recognizing that bearing witness to exploration means more than observing—it means understanding what's at stake, grappling with the questions the mission raises, and deciding whether the endeavor deserves your support.

Active witnessing starts with curiosity. Rather than accepting surface-level coverage that treats the mission as spectacle, dig into the technical details. Understand why the lofted entry profile matters, what proximity operations demonstrate, why life-support validation carries weight beyond the immediate mission. Read the NASA updates, follow technical journalists who explain rather than hype, ask questions when things confuse you. The mission becomes more meaningful when you understand what's actually happening and why each piece matters.

It extends to critical engagement. Ask whether the cost justifies the return. Question whether international partnerships are genuine collaboration or geopolitical theater. Wonder whether Gateway serves operational needs or primarily justifies continued spending. Honest questions strengthen support for exploration by forcing advocates to defend choices with evidence rather than rhetoric. Uncritical enthusiasm helps no one—not the program, which needs accountability to improve, and not you, whose support should be informed rather than reflexive.

Witnessing includes sharing what you learn. Talk with friends, family, students about what Artemis aims to accomplish. Correct misconceptions when you encounter them—no, the Moon landing wasn't faked; yes, commercial companies are involved but NASA leads; no, robotic missions can't fully replace humans for every objective. Help others understand why the mission matters beyond simplistic "it's cool" framing. The more people grasp what's at stake, the more resilient political support becomes when budgets tighten and priorities shift.

It means advocating when appropriate. If you believe space exploration deserves sustained funding, tell representatives. Write to congressional offices. Support science education initiatives that build the workforce future missions require. Vote for candidates who prioritize exploration alongside other values you hold. Political support for long-term programs like Artemis doesn't emerge spontaneously—it's built by citizens who communicate that these priorities matter to them.

And witnessing includes reflection on what exploration represents. When you watch the launch or follow the mission, consider what it says about human ambition, about our willingness to venture into the unknown despite risk and expense, about our capacity for cooperation across nations and generations. Consider what it means to live in an era when some humans travel beyond Earth while others still lack basic necessities.

Consider whether exploration drives progress or distracts from it. These aren't questions with simple answers, but grappling with them honestly makes you a more thoughtful witness.

The role you choose matters more than you might think. Large programs survive because enough people believe they should, and that belief must be refreshed continuously. Apollo collapsed partly because public support waned—not because people actively opposed it but because they stopped caring. Artemis faces the same vulnerability. If millions watch Artemis II and then forget about it, the program's fragility increases. If those millions remain engaged, ask questions, advocate when appropriate, and maintain interest through the gaps between launches, they help create the conditions for sustained exploration.

You're not a passive audience for a NASA production. You're a stakeholder in a collective endeavor that asks what humanity should pursue, what's worth the cost and risk, what future we want to build. Artemis II offers a moment to engage those questions seriously. What you do with that moment—whether you engage deeply or skim the surface, whether you ask hard questions or accept easy narratives, whether you stay involved or move on—shapes what comes next.

What Lies Ahead

When Wiseman, Glover, Koch, and Hansen climb out of the capsule after splashing down in the Pacific, they'll have validated systems that the entire Artemis program depends on. They'll have flown farther from

Earth than any humans in half a century. They'll have tested an entry profile born from engineering necessity that will protect all future crews. They'll have demonstrated that diverse, international teams can operate seamlessly in deep space. And they'll have provided NASA with data that makes Artemis III's lunar landing possible.

But validation isn't achievement. Proving that systems work doesn't automatically lead to sustained exploration. Political will must continue. Budgets must flow. International partners must remain committed. Technical challenges on subsequent missions must be solved successfully. The gap between Artemis II's success and the sustained lunar presence the program promises is filled with uncertainties that no amount of preparation eliminates.

What's clear is that the mission represents a threshold. Before Artemis II, returning humans to deep space remains aspirational—something NASA promises it can do, supported by analysis and uncrewed tests, but not yet proven with crew aboard. After Artemis II, if all goes well, it becomes operational capability—something demonstrated and repeatable, ready to support landing missions and Gateway operations. That threshold matters enormously. It's the difference between "we think we can" and "we know we can."

Whether crossing that threshold leads to sustained exploration or proves to be another brief surge before retreat depends on choices made in the years ahead—by Congress, by presidential administrations, by international partners, by engineers solving technical challenges, and by citizens deciding whether exploration deserves their support. No single constituency controls the outcome. The program's fate emerges from complex interactions among political forces, fiscal realities, technical capabilities, and cultural priorities.

For those watching Artemis II launch, the mission offers a moment to witness something that happens rarely: humans attempting what hasn't been done, pushing into territory that remains uncertain, accepting risk for reasons that include but transcend immediate practical returns. Whether that's worth $4.1 billion depends on your values and priorities. Whether it leads somewhere meaningful depends on what follows. But the attempt itself—the willingness to try despite expense and uncertainty—says something important about what humans are capable of when we commit to ambitious goals.

The countdown will reach zero. The engines will ignite. The vehicle will climb away from Earth, carrying four astronauts toward the Moon. What happens next unfolds across multiple dimensions—technical, political, symbolic, strategic—each influencing the others in ways that shape whether Artemis becomes the foundation for sustained exploration or another chapter in the long history of attempted and abandoned spaceflight programs. You'll be watching. What you make of it matters more than you might think.

Resources

You've reached the end of this book, but Artemis II's story continues to unfold. Some readers will want to follow the mission in real-time when it launches—tracking the countdown, watching splashdown, understanding what mission control's callouts mean. Others will want to dig deeper into technical details, reading NASA's engineering reports and primary documents. Still others will want to understand the political and budgetary context that shapes what comes next, or to verify claims made here against authoritative sources.

This chapter provides the resources for all those purposes. It's organized to help you find what you need whether you're a casual viewer wanting NASA's livestream or a researcher chasing down Congressional testimony about program costs. The sources listed here are authoritative, regularly updated, and represent the best available information on Artemis. Where opinions matter, journalism sources span different perspectives. Where facts matter, official documents provide ground truth. And where understanding matters, archives and technical documentation offer depth that no single book can match.

Think of this chapter as your toolkit for staying engaged with Artemis beyond reading about it. Bookmark the sites you'll use. Follow the social media accounts that match your interests. Download the documents that answer your questions. The mission belongs to everyone who chooses to pay attention— these resources help you do that effectively.

Following the Mission

Official NASA Sources

NASA Artemis Website: nasa.gov/artemis

The central hub for all Artemis information. Updated regularly with mission timelines, crew profiles, hardware status, and launch dates. When schedules slip or plans change, this is where official announcements appear first. The site includes educational resources, high-resolution images, and links to live coverage.

NASA TV: nasa.gov/nasatv

Live streaming coverage of launches, mission milestones, and press conferences. Available via the website, YouTube, and various streaming apps. Coverage typically begins four hours before major events with expert commentary explaining what's happening and why. During Artemis II, NASA TV will provide continuous updates from launch through splashdown.

Artemis Blog: blogs.nasa.gov/artemis

Mission control updates written by NASA's communication team. During active missions, the blog posts regularly—sometimes hourly during critical phases—with details on spacecraft status, crew activities, and upcoming milestones. More technical and comprehensive than social media posts but more accessible than engineering reports.

NASA Social Media

• Twitter/X: @NASA and @NASAArtemis

• Instagram: @nasa

• Facebook: facebook.com/NASA

Real-time updates, behind-the-scenes content, and mission highlights. Social media posts are designed for general audiences and provide the quickest way to get updates when news breaks. During Artemis II, expect multiple daily posts with photos, videos, and status updates.

Kennedy Space Center Visitor Complex: kennedyspacecenter.com

Information on viewing opportunities, tours, and educational programs. If you're planning to watch the launch in person, this site provides tickets, viewing area maps, and recommendations for accommodations. The complex also hosts mission exhibits and artifacts.

News and Commentary

NASA Press Kits: nasa.gov/artemis (under Media Resources)

Comprehensive pre-launch documents providing mission overview, crew biographies, hardware specifications, timeline details, and technical background. Released weeks before major missions, these press kits are invaluable for understanding what's about to happen. They're written for journalists but accessible to anyone willing to engage with technical content.

NASA Podcasts

• Houston We Have a Podcast: nasa.gov/podcasts

• Artemis-specific episodes featuring engineers, astronauts, and mission planners

Long-form audio interviews that provide depth beyond what press releases offer. Episodes typically run 30-60 minutes and cover everything from heat shield engineering to crew training to political challenges.

Research and Deep Dives

Primary Documents

Science Definition Report: Available through nasa.gov

The technical document outlining Artemis science objectives, mission requirements, and measurement priorities. Dense but authoritative—this is what NASA scientists wrote to justify the program's science value.

Artemis Accords: Full text at nasa.gov/specials/artemis-accords

The actual treaty language, not summaries. Essential reading for understanding international partnerships and the normative framework NASA is building. Includes explanatory notes on each principle.

Congressional Testimony and Budget Documents

Search "NASA budget justification" at nasa.gov/budget

Detailed breakdowns of how NASA requests and spends money, including line-item costs for SLS, Orion, Gateway, and related systems. Testimony from NASA leadership before Congressional committees provides rationale for budget requests and responds to oversight concerns.

Government Accountability Office Reports: gao.gov (search "Artemis" or "NASA human spaceflight")

Independent oversight assessments of program cost, schedule, and technical risk. GAO reports are critical but fact-based, identifying genuine problems without ideological bias. Essential reading for understanding legitimate concerns about the program.

NASA Office of Inspector General Reports: oig.nasa.gov

Similar to GAO but internal to NASA. The IG audits cost overruns, management practices, and contractor performance. Reports can be scathing but are rigorously documented. They provide accountability that's essential for public trust.

Technical Documentation

Space Launch System Specifications: nasa.gov/sls

Detailed technical data on the rocket's performance, including thrust profiles, propellant capacity, flight mechanics, and abort modes. Includes separate documents for Block 1, Block 1B, and future configurations.

Orion Quick Facts: nasa.gov/orion

Capsule specifications, life support details, avionics architecture, and European Service Module integration. More accessible than full engineering reports but still substantive.

Gateway Program Documentation: nasa.gov/gateway

Architecture plans, module specifications, orbit mechanics, and international contribution details. Regularly updated as the station design evolves and partners commit hardware.

Human Landing System Information: nasa.gov/humans-in-space/human-landing-system

Lander requirements, contractor proposals, mission profiles for surface operations. Includes both SpaceX's

Starship HLS and Blue Origin's Blue Moon designs.

Artemis III Landing Site Analysis: science.nasa.gov/moon

Scientific rationale for south pole targeting, site-specific geology, and resource assessment. These documents explain why certain craters and ridges matter more than others.

Reliable Journalism and Analysis

Eric Berger / Ars Technica: arstechnica.com/author/ericberger

Perhaps the most knowledgeable space journalist covering NASA. Berger combines deep technical expertise with access to insider sources. His reporting on Artemis costs, schedule challenges, and political dynamics is consistently ahead of mainstream coverage. Offers critical but fair analysis.

NASA Spaceflight: nasaspaceflight.com

Despite the name, not affiliated with NASA. An independent outlet with detailed technical coverage, breaking news, and insider reporting. The forums host discussions among aerospace professionals. Excellent for understanding hardware development and launch operations.

The Planetary Society: planetary.org

Advocacy organization founded by Carl Sagan that covers space exploration broadly. Their Artemis analysis tends toward supporting human spaceflight while advocating for robust science funding. Good for understanding how scientific community views the program.

Space.com: space.com

General audience coverage that's accessible without being superficial. Good entry point for readers who want more than headlines but aren't ready for technical journals.

Space Policy Online: spacepolicyonline.com

Covers the political, budgetary, and international relations aspects of space programs. Essential for understanding Congressional dynamics, international negotiations, and how policy decisions shape technical outcomes.

The Space Review: thespacereview.com

Weekly essays and analysis on space policy, history, and technology. Contributors include historians, engineers, policy analysts, and former NASA officials. Articles are long-form and substantive.

Historical Context

Apollo Mission Transcripts: history.nasa.gov/afj/index.html

Complete audio transcripts and mission logs from all Apollo flights. Reading these provides perspective on how different (and similar) deep-space operations were fifty years ago. The Apollo 8 transcripts are particularly relevant for comparison to Artemis II.

NASA History Publications: history.nasa.gov

Official histories of Mercury, Gemini, Apollo, Space

Shuttle, and ISS programs. These books—available free as PDFs—provide context on how major programs evolved, succeeded, or failed. The multi-volume Apollo history is essential for understanding why that program ended.

Constellation Program Archives: nasa.gov/constellation

Documentation of the cancelled predecessor to Artemis. Understanding why Constellation failed—budget overruns, technical challenges, political opposition—illuminates the challenges Artemis faces. Many Constellation concepts (Orion capsule, lunar south pole focus) carried forward into Artemis.

Space Shuttle Program Lessons Learned: nasa.gov/shuttle

Technical reports and operational experience from thirty years of shuttle flights. Many lessons—about cost control, safety culture, hardware reusability, international partnerships—apply directly to Artemis.

Essential Glossary

Abort modes: Procedures for safely ending a mission if critical failures occur during launch or flight. Different abort types apply at different phases (pad abort, abort to orbit, abort once around).

Apogee / Perigee: The highest and lowest points in an elliptical orbit around Earth. Apogee is farthest from Earth; perigee is closest. For lunar orbits, the terms are apolune and perilune.

Avcoat: The ablative heat shield material covering Orion's crew module. During reentry, Avcoat chars and erodes, carrying heat away from the spacecraft through chemical decomposition and mass loss.

Cryogenic propellant: Rocket fuel stored at extremely cold temperatures to keep it liquid. Liquid hydrogen (-423°F) and liquid oxygen (-297°F) are the most common. Requires insulated tanks and careful handling to prevent boil-off.

Deep Space Network (DSN): NASA's global network of large radio antennas that communicate with spacecraft beyond low Earth orbit. Stations in California, Spain, and Australia provide nearly continuous coverage as Earth rotates.

European Service Module (ESM): The cylindrical structure built by Airbus for the European Space Agency that attaches below Orion's crew module. Provides propulsion, power, thermal control, and consumables.

Flight Termination System (FTS): Safety mechanism that can destroy a rocket in flight if it veers off course and threatens populated areas. Consists of explosive charges on the core stage and boosters.

Free-return trajectory: A flight path that uses the Moon's gravity to return a spacecraft to Earth without requiring additional propulsion. Provides a safety margin if main engines fail after leaving Earth orbit.

G-force: Acceleration expressed as multiples of Earth's surface gravity. During launch, astronauts experience 3-4 Gs (feeling three to four times their normal weight). During reentry with the lofted profile, G-loads can reach 5-6 Gs.

Gateway: NASA's planned small space station in near-rectilinear halo orbit around the Moon. Will serve as staging point for lunar landers and research platform in deep space.

Hypergolic propellant: Rocket fuel that ignites on contact with its oxidizer, requiring no ignition system. Reliable but toxic. Used in Orion's service module thrusters.

ISRU (In-Situ Resource Utilization): Extracting and processing resources found on the Moon or Mars rather than carrying everything from Earth. Primarily focused on water ice that can be split into oxygen and hydrogen.

Launch window: The time period during which a rocket can launch and still achieve its intended trajectory. For lunar missions, windows are constrained by orbital mechanics and can be as short as a few minutes or as long as a few hours.

Lofted entry: The reentry profile Artemis II will use, featuring a single continuous descent through the atmosphere at a steeper angle than skip entry. Produces higher peak heating and G-loads but avoids the gas-trapping problem that affected Artemis I's heat shield.

Max-Q: The point during ascent when aerodynamic pressure on the rocket reaches its maximum, typically around 80 seconds after liftoff. Engines often throttle back slightly during Max-Q to reduce structural stress.

NRHO (Near-Rectilinear Halo Orbit): Gateway's unusual orbit, which takes seven days to complete and swings from close to the Moon to high above it. Gravitationally stable and requires minimal propellant to maintain.

Proximity operations: Maneuvers where one spacecraft approaches and flies near another, often in preparation for docking. Requires precise navigation and control. Artemis II's proximity ops demonstration tests manual flight capability around the spent upper stage.

Skip entry: The reentry technique used on Artemis I, where the spacecraft dips into the atmosphere, briefly skips back into space, then reenters fully. Extends landing range but creates thermal cycling that caused the heat shield gas-trapping issue.

SLS (Space Launch System): NASA's heavy-lift rocket designed for deep-space missions. Block 1 configuration used for early Artemis missions stands 322 feet tall and produces 8.8 million pounds of thrust.

TLI (Translunar Injection): The engine burn that accelerates a spacecraft from Earth orbit onto a trajectory toward the Moon. For Artemis II, TLI occurs roughly 28 hours after launch.

Van Allen radiation belts: Regions of intense radiation trapped by Earth's magnetic field. Spacecraft must pass through these belts to reach the Moon. Exposure time is brief but adds to total mission radiation dose.

For the Truly Committed

If you've read this entire book, followed the resources above, and still want more, consider these advanced sources:

Journal of Spacecraft and Rockets: AIAA's peer-reviewed journal covering aerospace engineering. Articles are highly technical but represent cutting-edge research on propulsion, structures, and mission design.

NASA Technical Reports Server: ntrs.nasa.gov

Searchable database of every technical report NASA has published. Thousands of documents on Artemis systems, testing, and analysis. You can lose weeks here.

Congressional Research Service Reports: crsreports. congress.gov

In-depth analysis prepared for Congress on space policy, budgets, and international competition. More accessible than GAO reports but still substantive.

Federal Aviation Administration Commercial Space Data: faa.gov/space

Licensing information and safety reports for commercial launch providers working with NASA, including SpaceX and Blue Origin.

ESA (European Space Agency) website: esa.int

European perspective on Artemis, with detailed documentation on the service module and other contributions.

JAXA (Japan Aerospace Exploration Agency): global. jaxa.jp

Japanese contributions to Artemis, including Gateway modules and lunar surface mobility.

Image Credits

Unless otherwise noted, all photographs courtesy of NASA and are in the public domain.

PHOTOGRAPHS

NASA Public Domain Images:
- Artemis II Mission Patch
- Orion's heat shield following Artemis I
- Cmdr Reid Wiseman inside International Space Station (2014)
- Victor Glover at Johnson Space Center (2022)
- Inside the Orion Crew Module
- Artemis I Flight Day 13: Orion, Earth, and Moon
- Earthrise

NASA Public Domain Images with Photographer Attribution:
- Kim Shiflett: Space Launch System (SLS) rocket and Orion spacecraft at Launch Pad 39B (March 18, 2022); Helga phantom torso at Kennedy Space Center; Orion off the coast of Baja California (Dec. 11, 2022)
- Glenn Benson: The Artemis flag at Kennedy Space Center, Florida (Jan. 6, 2023)
- Keegan Barber: NASA's SLS rocket carrying Orion spacecraft launching on Artemis I flight test (Nov. 16, 2022)
- Josh Valcarcel: NASA's Orion spacecraft splashing down in Pacific Ocean after 25.5 day mission
- Charles Beason: Artemis II astronauts Victor Glover, Reid Wiseman, Christina Koch, and Jeremy Hansen
- Daniel O'Neal: Canadian Space Agency astronaut

Jenni Gibbons practicing simulated lunar tasks at NASA's Neutral Buoyancy Laboratory, Houston
- Drew Morgan: View of the Gut on Chip CubeLab aboard International Space Station
- Jason Parrish: Artemis I Space Launch System and Orion spacecraft atop mobile launcher on Launch Pad 39B (Sept. 15, 2022)
- Joel Kowsky: NASA's Space Launch System rocket with Orion spacecraft illuminated by spotlights at Launch Complex 39B (March 18, 2022)

Creative Commons Licensed Images:
- Specialist Christina Koch aboard the ISS (2019) - Photo courtesy of NASA, CC BY 2.0
- Jeremy Hansen (2022) - Photo credit: Nattanon Dungsunenarn/Spaceth, CC BY 2.0

Author:
- Satellite Beach, November 2022 - Photo: Gray Sutton

GRAPHICS AND DIAGRAMS

NASA Graphics:
Diagram of NASA's SLS rocket carrying Orion spacecraft
NASA's RS-45 Engines infographic
Core Stage infographic
Orion Crew Module diagram
Artemis II Mission Map
Piloting Demonstration Test graphic

INFOGRAPHICS AND RENDERINGS

Microsoft Designer (AI-Assisted):
Artemis II Mission Timeline

"Following the Mission" infographic
Artemis Accords infographic
Skip-Entry Reentry Profile
Modified Lofted-Entry Reentry Profile
"What does a Billion Dollars Buy" infographic
Rendering of future base at Shackleton Crater
Rendering of Space Station Gateway
Rendering of Lunar Terrain Vehicle (LTV)

BOOK COVER

Cover design by Old Shrimp Road Press (AI-assisted using Microsoft Designer)

Thank you for spending your time on this book. If anything in these pages stayed with you—a line, a scene, a thought—I'd love to hear about it.

If you're willing, please leave an honest review. It's one of the simplest ways to support an author and help the right readers discover the story.

Scan the QR code to reach my review hub and choose your preferred site (Amazon, Goodreads, etc.).

Thank you—truly.

Gray Sutton

www.ingramcontent.com/pod-product-compliance
Lightning Source LLC
Chambersburg PA
CBHW061207220326
41597CB00015BA/1548